Policymaking for A Good Society:
The Social Fabric Matrix Approach to Policy Analysis and Program Evaluation

Policymaking for A Good Society:
The Social Fabric Matrix Approach to Policy Analysis and Program Evaluation

by

F. Gregory Hayden
University of Nebraska-Lincoln

 Springer

Library of Congress Control Number: 2005933285

ISBN:10: 0-387-29369-8   e-ISBN: 0-387-29370-1
ISBN-13: 978-0387-29369-1

Printed on acid-free paper

Printed in the United States of America.

9 8 7 6 5 4 3 2 1

springeronline.com

To Theresa M. Hayden

CONTENTS

List of Illustrations   ix

Preface   xiii

Chapters

1.   Introduction   1

2.   Policy Paradigms Should Be Consistent with the Complexity of Reality   13

3.   Instrumental Philosophy and Criteria   21

4.   General Systems Principles for Policy Analysis   51

5.   Social Criteria and Socioecological Indicators   61

6.   The Social Fabric Matrix   73

7.   Illustrations of the Social Fabric Matrix   109

8.   Timeliness as the Appropriate Concept of Time   145

9.   Evaluation for Sufficiency: Combining the Social Fabric Matrix and Instrumentalism   187

10.   The Social Fabric Matrix in a Metapolicymaking Context   199

Notes and References   229

Index   245

LIST OF ILLUSTRATIONS

Tables

3-1.   Alternative Models and Their Explanations of
       Different Phenomena   24

7-1.   Cost-Plus Percentages that Multiply CIC   119
       Costs

Figures

2-1.   Policy Analysis   17

2-2.   Integrated Systems   18

3-1.   Social Policy A: Antitrust Area   35

3-2.   Six Integrated Social Policy Areas   38

5-1.   Policy Analysis Paradigm   64

5-2.   Policy Analysis Paradigm with Primary
       Criteria   67

6-1.   Noncommon-Denominator Process Matrix   86

6-2.   Hypothetical Social Fabric Matrix   89

6-3.   Closed Digraph   91

6-4.   Unidirectional Digraph   91

6-5.   Simple Social Fabric Matrix   92

6-6.   Simple Social Fabric Matrix Digraph   93

6-7.   Data Management Spread Sheet   93

6-8.   Convergence of Balanced, Unidirectional, and
         Centralized Systems   96

7-1.   General Social Fabric Matrix of CIC System   111

7-2.   General Social Fabric Digraph Network of
         CIC System   112

7-3.   Social Fabric Matrix of Contracts and Costs
         during CIC Preoperational Phase   113

7-4.   Social Fabric Digraph Network of Contracts
         and Costs during CIC Preoperational Phase   115

7-5.   Social Fabric Matrix of Institutional
         Components during CIC Operational Phase   120

7-6.   High Level Mapping of CIC Network
         Structure   121

7-7.   CIC Operational Cost Network   124-125

7-8.   Social Fabric Matrix of the Daily Federal
         Funds Market   126

7-9.   Social Fabric Digraph of the Daily Federal
         Funds Market   129

7-10.   High Level Mapping of the Daily Federal
         Funds Market   130

7-11.   Digraph of the Open Market Desk System   131

7-12.   Digraph of the Federal Reserve System   132

7-13.   Social Fabric Matrix of Components that
        Generate Livelihood Strategies in
        Theethandapattu, Tamil Nadu, India   134

7-14.   SFM Digraph among Key Components of
        Livelihood Strategies   135

7-15.   General Social Fabric Matrix for Surface
        Water Management in Nebraska   138

7-16.   Social Fabric Matrix of the Current Era   139

7-17.   Digraph of Relationships among Regulatory
        Institutions under the 1984 Legislative Bill
        1106 in Nebraska   140

7-18.   Relationships among Regulatory Institutions
        and Water Users' Institutional Organization in
        the Application Process   142

8-1.   Evolution of Technology, Time Measurement
        Instruments, and Temporal Concepts   166

8-2.   Digraph of Overlapping Processes   173

8-3.   Process Digraph of Different Process
        Frequencies. High frequency of B is
        Modified for Low Frequency of A   179

8-4.   Evolution of Figure 8-2 with Acquisition and
        Loss of Nodes and Deliveries. Loss of
        Nodes Indicated by Shaded Circles. New
        Deliveries Indicated by Dotted Lines. Nodes
        Lost are 3, 9, and 10. Nodes Gained are 11,
        12, 13, 14, and 15.   181

9-1.   Wetland Ecosystem Model   190

9-2.   Delivery Criteria Indices (m and n)   193

9-3.   Decision Space with Two Criteria Indices   193

10-1.   Policy, Strategy, and Tactics of
        Policymaking: Phases and Levels of
        Metapolicymaking   200-201

10-2.   Bureaucratic Approach   203

10-3.   Pseudostrategic Approach   204

10-4.   Scholarly-King Approach   205

PREFACE

Society, ecological systems, and technological combinations are sets of ongoing processes that are organized as integrated systems and networks. Consequently, real-world problems—whether labeled social, economic, environmental, or technical—are a result of the ongoing processes that organize and coordinate integrated parts to make undesirable deliveries to each other. Furthermore, the processes are guided by numerous policies and concomitant rules, regulations, requirements, and enforced behavioral patterns. Therefore, there is no reason to expect processes to change or problems to be solved without policy changes. The processes are ongoing, so changes in undesirable deliveries are dependent on changes in policies.

One premise of this book is that too often policy analysis is conducted with knowledge bases and tools that are not appropriate for the task of analyzing and understanding complex socioecological and sociotechnical systems leading to wasted resources, policy failure, and frustration. The conjunction of the complexity of problem contexts and inappropriate policymaking that follows from insufficient analysis has left citizens frustrated and bewildered. Citizens want problems solved, yet they have lost faith in the ability of policymakers to implement solutions necessary to achieve a good society. Another premise is that it is not necessary to continue down that destructive path. In response, the purpose of this book, briefly stated, is to explain how to model, analyze, and make policy for the social fabric in which society's problems are enmeshed. Such policymaking requires that attention be given to the integration of an array of concerns including philosophy, engineering, institutional analysis, and ecology. The book includes that array of concerns—as well as others—but more important, it explains how the "social fabric matrix" can be used to design policy research, generate and analyze program scenarios, and conduct "metapolicy-making" consistent with problem resolution. The contribution expected from modern science and philosophy for solving problems has not materialized because of the absence of a general systems methodology for guiding cross-disciplinary research and policy analysis. Because policymaking is the most important inquiry a society conducts, philosophy, science, and systems theory need to be applied as an integrated whole. This book is about how to conduct such inquiry. Chapter 1 explains the general paradigmatic approach and how the concepts and concerns of the book are organized and elaborated in order to accomplish the book's purpose.

This book was completed with the assistance of an extensive sociotechnical system that provided me with research findings and assistance from colleagues, policy analysts in numerous kinds of agencies, scholars, research libraries, foundations, universities, consulting contracts, students, research assistants, professional associations, and so forth. Although it is not possible to individually acknowledge all the organizations involved and relevant persons in those organizations, their contributions are recognized as important.

The material in the book is very dependent on many different scientific and philosophical contributions made during the last 100 years. That recognition is not just perfunctory. Those contributions are recognized throughout the book as the foundational structure for the analysis, planning, and policymaking recommended. The book is very dependent on the scholars responsible for those contributions.

My service to governors, mayors, and legislators; involvement in policy research; consulting for various public, corporate, and non-profit organizations; and service on boards and commissions were crucial in the completion of research for and testing of the book's substance. Those experiences provided the opportunity to apply the social fabric matrix to real-world problems and to have the conclusions tested under fire in the advocacy process.

My students at the university and the research analysts in government agencies with which I have been associated have made two important contributions. First, they gave important feedback with regard to the effectiveness of the manuscript as it has been used as a learning text in the classroom and for research teams. Second, their research projects have demonstrated the instrumental practicality of the social fabric matrix approach.

Appreciation goes to Jan Hime, Amanda Mausbach, Kathleen Kramer, and John Hayden for the importation, creation, and programming of graphics to a common format for adding to the manuscript.

Pam Royal served as the project's administrative assistant, typed all the drafts of the manuscript, and worked tirelessly and cheerfully to see that the manuscript was finished. Her commitment and excellent work is especially appreciated.

By far, I am the most indebted to my wife, Theresa, to whom the book is dedicated. The book reflects her valuable ideas, constructive criticism, suggested improvements, and editing assistance. Additionally, her love and sustained encouragement have been crucial.

F. Gregory Hayden
Lincoln, Nebraska

CHAPTER 1

INTRODUCTION

The purpose of this book is to explain an approach to policy analysis and planning that will allow us to capture the complexity of the world around us and be consistent with modern science. We know the world to be very complex because modern social, physical, and ecological sciences have demonstrated the numerous, sometimes fragile, transactions among various system components and also because modern technology has created numerous, sometimes fragile, kinds of relationships among various aspects of the social, physical, and ecological world in which we live. The understanding we have gained from science and from experiencing the technological society on a day-to-day basis has led to a great transformation in our ideology and paradigm for viewing the world, but has yet to produce a good society.

Numerous books, articles, pamphlets, television shows, and courses of study have been devoted to explaining societal evolution and the paradigmatic shift that has transpired in the last 150 years. That work is taken as given, and this book is devoted to how to complete the analysis to make policy decisions in our private and public institutions and how to organize resources through policymaking that is consistent with our social beliefs and the needs of the natural environment. Because we no longer believe that life—as structured in an institutional and ecological milieu—is one-dimensional, our measures and analytical tools cannot be one-dimensional. Because we no longer think that beliefs and values can be ignored if, for example, we want successful irrigation systems or health care plans, an approach is needed to integrate what sociologists and anthropologists know about beliefs and values with the expertise of engineers, ecologists, agronomists, economists, physicians, and other expertise as needed for the problem at hand. This integration can no longer be the kind that has persons with different expertise working in isolation, and their independent work then placed under one cover. The analysts need to be guided by a common model or, to use Einstein's term, a common frame. The engineers' work must be guided by belief criteria, the sociologists' analysis should be consistent with the relevant technology, the economists' models need to be non-equilibrium systems, policymakers' actions are to be the result of integrated modeling, and so forth.

This book is based on a premise currently not in vogue in the public mind. The premise here is that we know enough, care enough, and have adequate resources and technology to solve our social, economic, and environmental problems. Or, stated differently, this book is optimistic by current standards of cynicism and pessimism. Our knowledge base is sufficient to do the research to understand our problems, our will is more than adequate, our work ethic is strong, our resources are abundant, and people are sufficiently educated to carry out the tasks in a technological society. One major deficiency is that we have not had the analytical means necessary to meld our will, knowledge, and institutions into a policy paradigm that allows us to obtain success.[1]

Current cynicism and pessimism regarding the possibility of success are not based on deficiencies of character, but, rather, on continual validation through failure. People work more hours only to watch their wages and salaries fall; they support more environmental protection endeavors only to learn that pollution levels continue to grow and species continue to be lost; people gain higher levels of education and training only to remain underemployed. Government employees work hard, gather more data than ever, and access greater computer capability, yet productivity problems continue to abound; tax payments to governments and consumer payments to corporations continue to grow, but every day we see infrastructure deteriorating and consumer products becoming shoddier. Production has grown on a global scale and trade has increased among nations, while per-capita income in over half of these nations is lower than a decade ago and slavery continues to grow and spread; the resource commitment to health care is massive, yet the reality of the American health care system is that it is very sick; and billions of public dollars are poured into farm payments, while commodity prices remain low and family farms continue to fail. In the 1999 edition of the book, *Environmental Protection: Law and Policy*, which is devoted to the synthesis of laws, regulations, and judicial doctrines for the legal profession, the authors state (1) that we no longer have "faith in the ability of officials to define and carry out policies that advance the public interest . . .,"[2] (2) that the fundamental tensions among science, ethics, and economics remain unresolved today, and (3) that fundamental social and environmental issues offer neither reconciliation nor peaceful resolution, but, rather, a set of contradictions which constantly defy solution.[3] For reasonable people, these observations validate a hypothesis of failure in the laboratory of reality. However, failure and contradiction are not necessary because it is possible to organize our resources and institutions in a manner consistent with the complexity of a modern society and in a manner to achieve a good society.

The traditional paradigm has crumbled. More correctly, it has finally crumbled after approximately 150 years of deterioration. Louis Menand explained in his recent book that it took nearly 50 years after the Civil War to develop the set of ideas to help people cope with the conditions of modern life. Those instrumentalist ideas were developed mainly by four people: Oliver Wendell Holmes, William James, Charles S. Peirce, and John Dewey. They changed the way Americans think, the views they express, and the way they live.[4] Now a new approach for socioeconomic and socioecological policymaking is required if the quality of our lives, our institutions in general, our political economy in particular, and the environment are not to continue to deteriorate. Karl Polanyi stated that the classical deterministic model was swept away with the massive broom of new legislation enacted around the world in the 1930s. As we look back, however, it is apparent that much of that legislation was allowed on an exceedingly experimental basis and was approved in desperate reaction to the global economic depression, not because of permanent changes in ideology or because of a paradigm shift for the community in general. That is not to say that there had not been a great loss of faith from the 1830s to the 1930s in deterministic models such as the classical unregulated market system. Yet the glimmer of hope that the faith could be resurrected continued to glow, at times rather brightly. In many ways, though, faith in the old paradigm died in the late 1960s and the 1970s when pronounced changes in ideological beliefs led to the realization, as George Lodge stated in 1974, that "the United States is in the midst of a great transformation, comparable to the one that ended medievalism . . . . The old ideas and assumptions are being eroded."[5] There is probably no better example of the emerging ideology during that period than the National Environmental Policy Act of 1969, because the act adopted an organic view for environmental protection. Much of the political ideology expressed since the 1970s can be interpreted as an attempt to restore the old ideology, and the reactionary policies formulated with the reinstatement of that old ideology have served the purpose of proving the reasons for its continued demise. This contemporary "testing" of reactionary hypotheses has unintentionally served the purpose of validating why the old approaches never made sense. The instrumentalist ideas which now guide Americans and the adoption of those ideas in the public arena have resulted in a difficult struggle against competing belief systems and powerful interests during the past 100 years. The adoption of instrumentalism, as developed by Holmes, James, Peirce, and Dewey has now emerged.[6] Instrumentalism forms the base of ideas upon which this book is built.

This book is neither about the demise of the old ideological view of the world nor about the rise of the new whole-system ideology that has been patched together from numerous sources in reaction to social and technological assaults on the traditional paradigm. Others have sufficiently logged that course and explained the emergence of the new view of the world in which we live. Now it is time for the development of new policy analysis and planning tools to successfully allow for a more systemic view to influence policymaking. The development of a new policy paradigm recognizes that relevant program evaluation and policy analysis must take into account the integrated system of beliefs that are relevant to the problem context and must recognize that any problem context contains different, and sometimes, conflicting systems of belief.

This is a "how to" book for policy analysts and policymakers to use for policy and program analysis in order to design policies and programs that more efficiently and effectively solve problems. It is a book about, to use Robert Westbrook's explanation of instrumentalism, how to conduct an appraisal through critical scrutiny by investigating the conditions under which problems arise and for estimating the consequences of acting upon them. These estimates are to consider the efficiency of the end-in-view for reconstructing the problematic situation. This involves "foreseeing the consequence of utilizing the means necessary to achieve the desired end and the consequences achieving it might have for the whole range of one's interests (that is, the way the end would itself function in subsequent experience as a resource or obstacle)."[7] To accomplish this, three concepts are dominant. They are context, criteria, and consequences. The policy tool kit must be able to (1) define the context that is producing the problem and the context that will exist after policies and programs are implemented to solve the problem, (2) apply criteria in order to judge which programs will achieve the desired ends, and (3) judge program efficiency by the consequences resulting from the policy actions.

During the last century, numerous books and articles have been written about what the desired ends of policy ought to be, the measurement of the common good, and the good society. Two books that were publication successes and received broad acclaim were books with the same title, *The Good Society*. One was written by Walter Lippmann in 1937[8] and the other by Robert N. Bellah et al. in 1991.[9] The success of both indicates that society continues to be concerned about policymaking for societal improvement. During the 54 years that separated their publication, society, in addition, became sensitive to the deterioration of our living systems and human health due to the intensive utilization of our ecological systems and the resulting pollution. In 1984, Herman Daly and John Cobb, Jr. recognized the growing environmental problem in their book, *For The Common Good*,[10] and explained that in order to be successful in understanding and solving a problem, policy analysts need to integrate social, economic, and environmental components. The need for such an integrated approach has been demonstrated by numerous writers during the last century. Examples in addition to those already mentioned include John Dewey's *The Public and Its Problems* (1927),[11] Karl Polanyi's *The Great Transformation* (1944),[12] Kenyon B. De Greene's *Sociotechnical Systems* (1973),[13] George Lodge's *The New American Ideology* (1974),[14] and Marc Tool's *The Discretionary Economy* (1979).[15] The intellectual, philosophical, and environmental base has been established, and grasped, by the reading public as necessary for establishing the good society. That base is taken as given. The purpose here is to build upon that base by explaining how ideas contained therein can be applied in real-world settings through the Social Fabric Matrix (SFM) to achieve the good society. The SFM is an integrated process matrix designed to express the attributes and relationships of the parts as well as the integrated process of the whole in order to define and appraise the real-world social, technological, and ecological system context that contains the problem of interest. Included in the context are the social, technological, and ecological criteria that can also be articulated and integrated in the SFM. Social life is a series of transactions among institutions, technology, persons, agencies, and the elements of the ecological system. In reality, these components are guided by the application and enforcement of normative criteria. To understand social problems, the criteria need to be included in the policy analysis model. The diverse array of social, legal, ecological, economic, financial, technological, and belief criteria that help shape and structure the institutional context can be included in the SFM to determine their importance and to determine whether the criteria may need to be changed in order to solve socioecological problems. Furthermore, the SFM allows for the identification and measurement of the consequences of policy and program implementation.

The book will occasionally note the failures of approaches traditionally used for evaluation and analysis. For example, failures of neoclassical cost-benefit analysis are mentioned, along with its failure to pass judicial muster. Given its failures, cost-benefit analysis will be less and less relied upon to guide real-world policymaking in the future. The main purpose of the book, however, is not to critique other approaches. It is, instead, to develop and explain a policymaking approach for solving problems. That approach is explained in Chapters 2 through 10.

Policy Paradigms Should Be
Consistent with the Complexity of Reality

The general paradigmatic approach and its components are introduced through two simple diagrams in Chapter 2 (Figures 2-1 and 2-2). The first exhibits the process of policy analysis. The second outlines the relationships among the components of a socioeconomic system. The components are (1) cultural values, (2) social beliefs, (3) personal attitudes, (4) social institutions, (5) technology, and (6) the ecological system. All these components need to be integrated to understand a problem context and to plan policy to solve a problem because the components are not separate in reality. They are unitary in the configuration and integration of our institutions such as business corporations, families, government agencies, hospitals, and so forth.

Instrumental Philosophy and Criteria

Scientific analysis is laden with beliefs and values; therefore, analysts and social thinkers in general should explicitly state the philosophy and criteria guiding their scientific work. Chapter 3 does so. The philosophy being followed is that laid down by instrumentalists such as

Charles Peirce, John Dewey, Rollo Handy, and Richard Mattessich. Attention to both scientific and social criteria is crucial in undertaking policy analysis and planning. The selection and handling of criteria depends on the philosophy adopted. The philosophy that was influenced by and is consistent with democratic policymaking is designated as instrumentalism. Three conceptual ideas guiding instrumental policymaking are (1) the transactional approach to science, (2) a problem orientation, and (3) judging results by consequences. Their importance for policy analysis is explained in Chapter 3.

General Systems Principles for Policy Analysis

In terms of analytical approaches to systems, most of the principles of general systems analysis (GSA) are consistent with instrumental policy analysis because GSA is a body of principles that are formulated to be relevant to all systems whether social, technological, ecological, or economic. The development of instrumentalism has been concurrent with the development of GSA. Although both areas were inspired by the recognition of transactional processes and holistic systems, very little intellectual cross fertilization and integration of knowledge bases has transpired between the two intellectual traditions. Through the years, knowledge, scientific methodology, and experience accumulated across scientific disciplines that converged to find commonality among systems. In Chapter 4, twelve principles of systems theory that are consistent with instrumentalism are explained. Policy analysis approaches that attempt to deal with the world in a systemic manner should be consistent with those twelve principles.

Social Criteria and Socioeconomic Indicators

Chapter 5 builds on instrumental philosophy and system principles in order to establish an approach for measurement and indicators. The approach emphasizes that the creation of measurement and indicators is socially guided; therefore, the social context, beliefs, and values must be defined early in the process of determining indicators, and interjected throughout the process of arriving at numerical representation.

Numerical indicators and measures are very abstract. Their generation and collection begin with basic values and beliefs that help define a social problem for which indicators are needed and with social beliefs about how to solve the problem. From that base, measurement theory is applied, survey and collection procedures are designed, and data collection is undertaken. Throughout the process, criteria are designed and applied to arrive at measures and indicators. After techniques are stacked on procedures which are stacked on criteria through assumptions and more techniques, numerical representation is completed. Thus, numbers, that is, numerical facts upon which we must be dependent for policymaking, are abstractions from the philosophy and social criteria used to create them. Numerical indicators are one of the most abstract entities of the research process. These facts are laden with values, beliefs, and ethical standards. That is why they can be so useful. Yet, they cannot be useful without the technical means for their use. The SFM, as explained in Chapter 6, is a means to organize facts for meaningful policy analysis.

The Social Fabric Matrix

Although policy science must be abstract, policy research and analysis must be concretized for a particular context with particular beliefs, institutional, technological, and ecological entities; and with the specification of relationships among them. The SFM, which is explained in Chapter 6, is the tool kit for articulating and integrating various categories of concern for the study of real-world contexts. One set of concerns includes those of (1) philosophy, (2) theory, (3) statistical and mathematical techniques, and (4) policy. The SFM allows for the expression of a philosophical context that is normative, deontological, and systemic. It is constructed to allow for the expression of theories and principles and to encourage the collection of data and indicators in a manner to both test and utilize those theories for policy. Another set of concerns is drawn from anthropology, social psychology, economics, and ecology. As mentioned above, they are (1) cultural values, (2) societal beliefs, (3) personal attitudes, (4) social institutions, (5) technology, and (6) the ecological system. They serve as the main components of the SFM. Criteria, institutions, and ecological systems regularly undergo evolutionary transformations. The SFM expresses these concerns by articulating the integration of flows and deliveries.

The cells in the SFM are not determined by mathematical techniques, but, rather, are determined by what is found in reality. Yet, the SFM allows for the application of a broad and diverse range of statistical and mathematical techniques, especially Boolean techniques. The SFM is designed for policy analysis. This is so because it allows for the description of a "what is" matrix in order to define the problem, as well as a comparative "what ought to be" matrix of the policy and programs that are under consideration in order to compare the current state of affairs to the state of affairs expected from the new programs.

The relationships and components of a socioecological system call for an array of different kinds of measures and indicators in order to define and evaluate a system. This array can be handled by the SFM; thus, it contains the database necessary for the evaluation of policies. The SFM can be used to measure the efficiency and effectiveness of policies by defining the system before the application of new policies and comparing the pre-policy SFM to the post-policy SFM. In summary, the purpose of the SFM is (1) to conduct policy analysis in a manner to understand the multidisciplinary system that produces the social, economic, ecological, or technological problem; (2) to design and analyze an alternative system based on policies under consideration in order to determine the consequences to expect from the policies; and (3) to compare the consequences of the original system with the consequences of the alternative system to determine whether a policy set is one that improves or worsens overall conditions when the array of impacts is compared for the systems.

Illustrations of the Social Fabric Matrix

Chapter 7 is devoted to the presentation of four different abridged SFM and digraph studies in order to illustrate the systems derived from the SFM approach and the diverse kinds of problems and policy contexts to which it can be applied. The first two cases presented also demonstrate the use of the computer program, *ithink,* which is a program that is useful for completing the digraph and processing data for the SFM.

Timeliness as the Appropriate
Concept of Time

Modern concepts of time are one of the most ignored issues of importance in policy analysis. Modern time concepts are implicitly ignored to the detriment of the analysis when time stream discounting techniques are used to determine value. Modern time concepts need to be brought to bear for the analysis to be useful for policymakers. In modern thought, time is no longer an exogenous concept, but, rather, another element in the sociotechnical system. Different temporal conditions occur for different kinds of institutional experiences. There can be a difference in temporal rhythms and temporal clocks from institution to institution. One time and one clock do not exist across institutions. Time changes among institutions, especially in a complex society. For policymaking to successfully sequence and coordinate the social deliveries of programs consistent with when deliveries are needed, the concept of "timeliness" should guide policy analysis. Timeliness is based on the idea that the correct amount of social goods and services be delivered at the appropriate point in the social process for the integration, maintenance, and improvement of the system to which policies are directed.

Chapter 8 will explain modern concepts of time, the importance of timeliness, and how the SFM and digraph can be utilized to conduct time analysis and to plan for timely delivery. Changes in the substantive flows and deliveries in the SFM and digraph also provide for real-time systems. Traditional time concepts and clocks are not sufficient for the space-time coordination that is needed to solve social and ecological problems. The timing system needs to internalize events in a socially relevant sequence. Such sequencing places social problems into a context which is ideally timed by the succession of events and deliveries relevant to that socioecological context.

Evaluation for System Sufficiency: Combining the
Social Fabric Matrix and Instrumentalism

The purpose of Chapter 9 is to combine the concepts of the SFM and instrumentalism with the knowledge of complex social systems in order to explain evaluation for system sufficiency. Furthermore, the chapter explains why concepts from the discipline of ecology such as "sustainability" and "biodiversity" are neither meaningful nor operative except in the context of sociotechnical systems.

The Social Fabric Matrix in a Metapolicymaking Context

The purpose of the final chapter is to provide an overall view of the policymaking of policymaking, that is, "metapolicymaking," and to place the SFM approach to policy analysis in the context of a metapolicymaking paradigm. The metapolicymaking paradigm includes the phases of policymaking, ranging from theory and philosophy to advocacy and budgeting, and the phases are divided among levels of policymaking—policy, strategy, and tactics. The phases and levels are explained in conjunction with each other, emphasizing that a systemic approach to research like the SFM is necessary for successful policy to be formulated and made operative.

CHAPTER 2

POLICY PARADIGMS SHOULD BE CONSISTENT WITH THE COMPLEXITY OF REALITY

There is a plethora of policy analysts, research consultants, professors, planners, and scientists with concomitant labs, bureaus, university research departments, policy institutes, survey centers, and so forth. Coincidently, the world's problems continue to grow. Yet, the recommendation here is that more research is needed. E. F. Schumacher, on occasion, stated that a neurotic is one who, upon discovering that he is going in the wrong direction, doubles his speed. Is the thesis of this book based on such neurotic tendencies? No. Evidence abounds to demonstrate that some research has been going in the wrong direction. Arguments, however, do not follow that encourage a continuation of that journey. The argument is that a holistic, integrated, and systemic methodology is needed if we want to avoid wasting our research resources, or, worse, creating more serious problems. Research has been used as a powerful weapon to support policies that help create severe environmental and health problems.[1] Such research is not completed by analysts trying to improve the quality of life and the environment. The more serious problem is the case of researchers who are well meaning, yet have a very narrow or misguided concept of reality, and use scientific models consistent with that concept.

   For example, some neoclassical economists occasionally startle the public by recommending the selling of babies, encouraging smoking because it kills people, approving the institution of women selling themselves into slavery, and similar kinds of policy conclusions. At a professional economics meeting I attended, a woman economist accused her male colleagues of being vicious because they approved of women selling themselves into slavery. That accusation, however, is a mischaracterization if one understands economics as it is often espoused and taught in university economics departments. A component of the model which commands the brain ware of many economists is an assumption that what people do in everyday life is determined by utility maximization of the participating individuals. If that is the model in one's tool kit, then it follows from that model that women who sell themselves into slavery have rationally calculated their utility functions, and they are selling themselves because it gives them the greatest satisfaction. Using classical logic, this is an acceptable conclusion given the model. The gender and personality traits of economists are not the issue, however. The problem is with the utility maximization model of analysis.

First, utility does not exist in the real world; thus, people cannot calculate utils or maximize utility. The idea of utility was found to be invalid by social scientists in the 1800s. Second, extensive use of hedonistic pleasure seeking as a determinant of behavior is dangerous and has been rejected by all societies. Third, utilitarian theories, although verbally adopted by authoritarian collectives like the Nazi powers of the World War II era, cannot guide policy because they are neither internally consistent nor operational. Fourth, the idea of utility maximization is based on the idea of action and bargaining among atomistic individuals in the marketplace, while the real-world effort of policy needs to be based on how to reach a reasonable consensus among overlapping institutional organizations. The idea that consumption in the market could make social relationships transparent is an illusion, "an illusion of transparency, an illusion of readable social relations, behind which the real structure of production and real social relationships remain illegible."[2] Finally, the utility maximization which neoclassical economists claim is captured by market procedures emphasizes a given fixed procedure for decision making rather than a focus on outcomes and how to change procedures to achieve desired outcomes. Policymaking encompasses outcomes as well as procedures and societal as well as individual concerns. Ideas like utility maximization ignore culture, social beliefs, institutions, power relations, traditions, procedures, and so forth, and, therefore, are not useful with regard to real-world policy analysis and decision making. Changing rules and procedures to implement social beliefs about what will make for a good society is usually an important part of policymaking. "To realize justice in its fullest sense—as encompassing outcomes as well as procedures and societal as well as individual consideration—it seems that a shared conception of the good *is* necessary."[3] Iris Young found with regard to the controversy surrounding the siting of hazardous waste sites that much of the controversy revolved around fixed procedures. "The issue of justice raised by community residents in the siting case, however, calls into question just those institutional structures that justify some decision making procedures."[4] Consequently, utilitarian ideas for capturing and analyzing the real world are irrelevant to those who want reliable policy analysis.

Neoclassical economists who assume that hedonism is an appropriate base for making decisions and that utility exists further assume that the pecuniary prices charged by corporations can serve as a reflective measure of utility maximization. Thus, they argue that corporate prices can be utilized as the measure of benefits and costs for the analysis of public programs and that monetary prices are to be the common denominator. This approach creates serious political problems for public policymaking because it enhances the measure used and control of analytical outcomes by those organizations whose interests, transactions, and decisions are expressed in monetary terms—that is, corporate organizations. This is true for most of the policymaking concepts recommended by the neoclassical paradigm such as cost-benefit analysis, the Coase theorem, and Pareto optimality.

The adoption of price as the measure of value endows corporations with exaggerated legitimacy and power in three ways. First, in a semiotic sense, corporate symbols are elevated to serve as the standard for policy analysis, and, therefore, the legitimacy of the corporate organization is elevated. To take the symbols of one institution as the measure and purported common denominator of a complex social process alienates and demotes other symbols in the minds of citizens and policymakers. If market prices are elevated to serve as the common denominator for everything, where does this kind of measure leave religious organizations, for example, who must determine the number of children who can be fed by a government program? Policymakers will inquire about the *dollar* value of feeding children. How do environmental organizations calculate the number of species that can be saved by a program that has been advocated? What is the dollar value of an endangered beetle in a South Dakota wetland? Emphasis on the monetary symbol of corporations as the correct measure shuts out the measures of other institutions such as family, religion, government, NGO, courts, science, and so forth. The other institutions are limited in making their case about concerns and criteria that are important to them. Second, in terms of political power, the selection of the corporation's criterion of success—that is dollar flows—as the social criterion of success provides a definite advantage in terms of political legitimacy, standing, and power. Whatever increases dollar flows for corporations is measured as an increase in social welfare. Third, the selection of corporate price as the appropriate measure means the analytical apparatus of the corporation becomes the dominant model for analysis. Financial accountants of corporations, for example, are experts on corporate dollar flow and, therefore, they possess the expertise to dominate analysis and discussion in the policymaking process. Religious leaders, welfare mothers, and ecologists, on the other hand, are not experts in financial discounting and cost accounting, and, thus, are at a disadvantage in the process. How is it possible to argue for clean air to prevent asthma in children when limited to market prices and dollars as the criteria?

In terms of a policymaking paradigm, the argument that price is a measure of social value is what logicians term a "category mistake." A category mistake is the treatment of a concept as if it really belongs to one logical type of category, when it belongs to another. For example, to say that the square root of four is white makes no sense because it is impossible meaningfully to predicate the color of a number. When an economist argues that the price an individual is willing to pay a corporation is the criterion for judging value, a category mistake is made. The willingness to pay for a good or service is a subjective want. That is inconsistent with the purpose of public policy. The purpose of public policy is to provide for social beliefs through political association and public processes. When asking for public programs to submit to the criterion of market price, "the economist asks of objective beliefs a question that is appropriate only to subjective wants. . . . One cannot establish the validity of these beliefs by pricing them, nor can that mechanism measure their importance to society as a whole."[5]

Approach to Policy Analysis and Evaluation

Figure 2-1 can be used to assist in understanding a more realistic approach to policy analysis and evaluation—one might say, an approach cluttered with the complexity of reality. Figure 2-1 is a schematic representation of a policy analysis paradigm that follows the lead of the policy scientist Yehezkel Dror[6] for designing indicators intended to serve the purposes of public policy. Figure 2-1 indicates, starting on the left, that social beliefs, values, and ethical standards are prerequisites for determining social goals and establishing primary criteria. Pri-

Figure 2-1. Policy Analysis

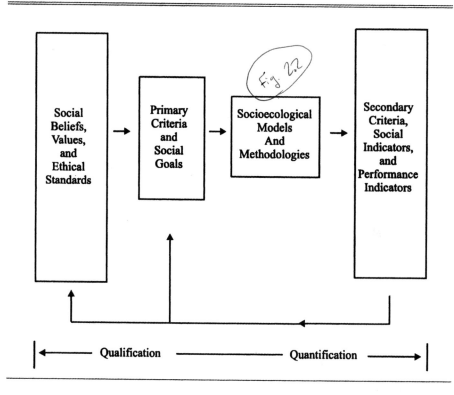

mary criteria are put into operation and monitored through the development of secondary criteria.   Secondary criteria are the social and performance indicators or measures. Consistent with Instrumentalism, Figure 2-1 reflects John Dewey's concept of social measurement as a spectrum from qualification to quantification.[7]   It includes a "feedback loop" from the secondary indicators back to social beliefs, ethical standards, and primary criteria in order to reflect that in public policy-making, the secondary indicators will provide negative or positive information feedback to those entities.

Secondary criteria, or measurement indicators, are found through the application of socioecological models and methodologies. The adequacy of the modeling will determine whether the indicators and measures have any meaning and relevance to societal goals.   The examples above about slavery and selling babies are cases where societal beliefs and goals are not being integrated with secondary criteria, modeling, and indicator creation.

Design and Application of Socioecological Models

The design and application of socioecological models should be completed in a manner to demonstrate and explain the relationships among the components relevant to the problem being studied. As outlined in Chapter 1, the components to be integrated in order to understand a system are: (1) cultural values, (2) social beliefs, (3) personal attitudes, (4) social institutions, (5) technology, and (6) the ecological system. All these components should be integrated to understand a problem area or to plan policy to solve a problem because they are not separated in reality. Figure 2-2 is an illustration of the relationships among the components that will be explained more fully below in Chapter 6. As is apparent, the components function together as a system because of the deliveries and flows among the components.

Figure 2-2. Integrated Systems

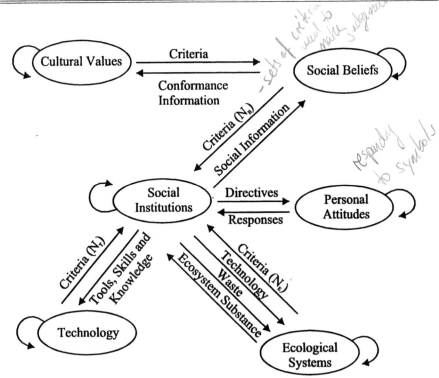

Figure 2-2 is in marked contrast with much of the modeling used in policymaking. Ecological system models which contain no social institutions or technology and only a measure of one flow within the system are common in ecology literature. Examples are models for wetlands located in rural areas that do not include the institutions and technology of agriculture and that include only energy measures among components in the wetlands. However, it is the practices of agriculture that set in motion the sociotechnical processes that deliver eroded soil to the wetlands, soil that fills the wetlands and carries pesticides and herbicides that kill species in the wetlands. The literature in economics regarding economic production models are often equally naïve. Such models usually are based on the Cobb-Douglas production function. On the input side, Cobb-Douglas does not include natural resources from an ecological system, entrepreneurial ability, energy, social beliefs, technology, financial capital, and so forth. On the output side, the production function does not include pollution, although the production of goods and services is not possible without pollution. Such ecological and economic modeling is not useful for policy analysis to solve problems of degraded ecological systems.

The normative criteria illustrated in Figure 2-2 are most important entities for understanding and analyzing any policy concern. The three normative sets of criteria are social belief criteria (N_B), technological criteria (N_T), and ecological system criteria (N_E). All three are criteria for the judgment of social institutions. Social beliefs are expressed in legal statutes, contracts, agency rules, regulations, operating procedures, and legislation.

N_T and N_E are not defined in an anthropocentric sense. Technology does not think about and decide upon normative criteria. Technology is the combination of tools, skills, and knowledge that is employed by social institutions such as corporations. The technological norms are the criteria conveyed to society as a result of the combination selected by particular societal units. Once selected and adopted, technology becomes woven into the social fabric "in such a fashion as to build its own necessity."[8] Likewise, no assumption is being made that an ecological system designs beliefs from mental reflection. Instead, N_E represents the normative criteria consistent with the maintenance of a particular kind of ecological system as institutions apply technology to the ecosystem in order to extract resources and dispose of waste. To change institutionalized waste disposal systems, for example, policymakers need to change the set of criteria, N_E, used to judge the systems.

It is through institutional structures and patterns that the various normative criteria are expressed. Likewise, institutions are the battleground for the clash among various criteria. For example, there is a traditional social belief (N_B) that workers' health should be protected, yet a new technology may be implemented that requires a new technological norm (N_T) which produces a certain level of cancer among the workers. Environmental impact analysis submitted by corporations as part of a licensing process for a particular technological design for waste disposal is required to include projections for the expected level of cancer. The policy process will decide which set of criteria is to be applied to evaluate the institutional and technological structure. Or, as a different example, people may decide they would like to alter the current ecosystem by implementing new criteria for less hazardous waste to be delivered from corporations to groundwater. This would initiate a conflict between the new criteria and current technological criteria (N_T).

The normative criteria are necessary for a social system to establish efficiency. N_B, N_E, and N_T deliver sets of subcriteria to authority institutions such as courts. The designated subcriteria become the standards to be applied by the institutional authority to determine efficiency. Efficiency means the ability to produce or achieve a desired effect. For society, the desired effects are determined by the normative criteria which are utilized to evaluate and judge efficiency. For example, courts deliver authoritative codes and regulations to corporations in order to establish the obligations of corporations with regard to requirements to protect the ecosystem. In turn, the corporations that are given authoritative power deliver requirements to processing units such as factories. Such requirements, for example, deal with supervising, auditing, monitoring, producing, storing, loading, transporting, and disposing of hazardous waste. For efficiency, the requirements must be enforced consistent with normative criteria.

Given the importance of instrumental philosophy and normative criteria in policymaking, the next chapter is devoted to explaining their relationship to the policy context.

CHAPTER 3

INSTRUMENTAL PHILOSOPHY AND CRITERIA

The philosophy whose development was influenced by and is consistent with democratic policymaking is designated as instrumentalism, transactionalism, or pragmatism. Three conceptual ideas from instrumentalism stand out as especially important for guiding policymaking endeavors. They are: (1) the transactional approach to science, (2) a problem orientation, and (3) judging by consequences. A transactional approach defines how we should go about observing and studying phenomena if we are to successfully gain an understanding of how the phenomena function. The transactional approach designs observation processes in order to observe and understand a problem within a full ongoing systemic process. A problem orientation requires that the problem itself should guide what kind of system should be designed to collect data, gain comprehension, and formulate policy. Instrumentalism differs from other philosophies in emphasizing that policy decisions and evaluation ought to be based on consequences, broadly defined. Instrumentalism is, of course, much broader than these three concepts, including ideational definitions for democracy, intelligence, association, community, scientific criteria, facts, and so forth. The main concern of instrumentalism is to provide assistance to the community for problem solving. This concern can be simply stated, as follows:

(1) Humans spend their lives in organizational relationships and cannot function except through viable association. The organizational relationships overlap, intersect, and interact in a complex interplay of discourse and action.

(2) Humans cannot experience viable association except through common beliefs and symbols, thus, association itself does not constitute a community. Productive association requires common beliefs, common symbols, and obligations that are mutually recognized.

(3) Common beliefs, symbols, and recognized obligations cannot exist except through common education and socialization processes.

(4) Effective and viable education and socialization processes are not possible except through inquiry into how to structure such processes.

(5) Inquiry must be constant and ongoing if social processes are to be improved and problems solved.

(6) An ongoing inquiry process cannot exist except through community institutions and organizations.

(7) To undertake the inquiry necessary for ensuring viable community institutions and problem solving, a transactional approach to inquiry is necessary.

A society includes obligations, rules, and requirements which require decision makers to determine socially appropriate actions.[1] This requires policymakers to complete efficiency evaluations in order to make judgments based on criteria drawn from social beliefs. Thus, efficiency considerations in policymaking begin with social criteria in order to know what is desirable. People living together in a social system requires that members be immersed in obligation to one another. In a social system, policy choices and actions are about institutional and organizational actions. "Action is behavior under the governance of some larger understanding . . . ."[2]

Transactional Approach of Instrumental Inquiry

The transactional approach to inquiry may be best understood by contrasting it with the approaches of self-actional and interactional approaches.[3] The self-actional model is a means of inquiry where things are viewed as acting under their own powers. Explanations of phenomena are attributed to essential objects such as great leaders, germs, rational consumers, superior genes, and so forth. In the self-actional approach, humans are usually given superior status. As Stephen Jay Gould explained, it is as if evolution were carried on for four and a half billion years in order to establish that the superior human being is the most important component of all processes.

When the self-actional model is complicated to the point of indicating relationships, a system is perceived to be made up of atomistic agents bouncing off each other as they undertake their self-interested action. Self-guided action radiates out from each individual and independent part. The first row of Table 3-1 indicates the kinds of self-actional explanations given with regard to intelligence quotient (I.Q.) scores, hunting, and finance. To explain high I.Q. scores, the self-actional proponent would attribute it to the child's being smart; to explain hunting success, a myth would be created about a great hunter, for example, the myth surrounding the American fur trader Jim Bridger; and, to explain financial success, the proponent of such modeling would attribute it to financial genius.

Further observation of systems brought the realization that more than individual agents were involved in any particular situation, thus, encouraging the adoption of interactional models. The interactional model emphasizes that, for example, across from the financial genius were stock sellers from whom the genius was buying, or bond buyers to whom sales were being made. Likewise, in the situational equation with the great hunter was the hunted, and with the smart student was the involvement of a teacher.

The interactional model followed the mechanistic balance theories of Newtonian physics, where thing is balanced against thing in a causal interconnection. "Inter" means between, and the emphasis was on the relationship between agents. This led to numerous kinds of equilibrium models, such as the equilibrium between supply and demand in economics.[4] Often a moral or value connotation was given to the equilibrium, indicating that it was good. Not only was it an equilibrium from which we should not expect deviation, but one from which we should not want deviation. Returning to Table 3-1, the second row contains interactional explanations.

The interactional explanation for I.Q. scores became the relationship between the student and the instrument, or between the student and the teacher. Considerable resources were devoted to studying the latter relationship. These studies, referred to as "Pygmalion in the Classroom," explained that when teachers were led (by the investigator) to believe a student was really very bright, the student's scores improved markedly on the next I.Q. exam because the teachers treated the

Table 3-1: Alternative Models and Their Explanations of Different Phenomena

Concerns / Models	Intelligence Quotient	Hunting of Game	Finance
Self-actional Model	Smart child.	Great hunter.	Financial genius.
Interactional Model	Interaction between student and teacher or between student and test instrument.	Match, check, and balance between cunning of hunter and prey across geographical area.	Equilibrium of supply and demand.
Transactional Model	Overlapping systems of school quality, home life, teachers' concern, early learning opportunities, school board policy, sales taxes, federal laws, nutrition, court decisions, life experiences, productivity of the economy, beliefs, attitudes, family income, activities of federal, state and local governmental agencies, and so forth.	Overlapping systems of federal, state and local governmental agencies, National Rifle Association, Scout organizations, property rights, court decisions, Congress, Agriculture Stabilization and Conservation Act, Environmental Protection Act, Natural Resource Districts, environmental advocacy groups, gun clubs, gun and ammunition producers and distributors, gun importers, hide processors, and so forth.	Overlapping systems of banks, Federal Reserve System, G-8, International Monetary Fund, World Bank, international electronic currency, saving and loan associations, Resolution Trust Fund, FDIC, Community Reinvestment Act, regulations of federal, state and local governments, property rights, court decisions, government and labor union pension funds, comptroller of currency, stock and bond markets, and so forth.

student as a bright student. Likewise, if teachers were led to believe that a student with high scores had low scores, teachers no longer gave him the attention of a gifted student, thereby causing scores to fall.

The explanation for hunting prowess, under the interaction model, was a game of "cat and mouse" between the hunter and prey as they matched and checked (balanced) each other with attacks and counter moves across a geographic area. This mythical model was impressed into the popular mind at its most basic level in the movie *Sergeant York,* in which Gary Cooper showed the geometric balance in turkey hunting which was supposed to have served him well later in similar action in war. The interaction balance theory for finance was explained as the traditional equilibrium between the forces of supply and demand.

The self-actional and interactional models appear rather simplistic when confronted with the complexity of reality. This is the reason for the evolution of scientific modeling to the transactional approach. "Trans" means across, and the emphasis is on the reality that there are numerous rules, regulatory criteria, enforcement agencies, laws, institutions, and beliefs across any relationship or transaction; numerous overlapping forces guide the agents and their actions. For example, a can of Coca-Cola may have the same properties for the chemist whether it is manufactured and distributed in an Islamic country or a capitalist country. The beliefs, property rights, and decision making rules that guide the manufacture and distribution, however, are very different, thereby leading to different production structures, distribution patterns, and pollution levels. The Coke, as it is passed among the social transactions, does not function by itself. It is not self-actional. Neither are human agents and institutions that are connected to Coke.

The transactional approach is a system of theories explaining action without attributing final motive to elements; other detachable, reduced, independent entities, agents, or to reality.[5] Neither is an attempt made to detach relations from the elements or agents. Transactional analysis was pioneered in physics, and emphasizes fields, within fields, within fields. It is even more valid in the social and ecological sciences and is usually applied through integrated process analysis. A transaction can be taken as designating the full ongoing process in a field where the connections among the aspects and phases of the field and the inquirer are in a common process.[6] The use of the word "common" is not to imply harmonious. A common process may exhibit conflict, violence, and destruction as a regular characteristic of the process.

If we think about studies completed with regard to the three areas in Table 3-1, their transactional complexity is apparent.   With regard to I.Q. scores, we know that home life, school quality, teacher concern, early learning opportunities, property taxes, federal law, nutrition, court decisions, life experiences, productivity of the economy, beliefs, attitudes, pollution, family income, and a whole host of other transactional patterns are involved in determining the kind of I.Q. score received by a student.   This means that I.Q. scores are rather arbitrary as a measure of student ability, and thus courts, in the United States, have become involved to guide their use and in some cases outlaw the tests completely.   Again, the court is playing the role of one of the many institutions that are across the relationship between students and the I.Q. instrument.

When we study the so-called great hunters of the past, we find hunting expeditions, not self-actional loners.   And seldom have the social hunting organizations in U.S. history confronted nature through an interactional balance.   In fact, the opposite has been the case.   For example, after the organized buffalo hunters finished systematically applying their technology, the buffalo were gone.   The myths surrounding Jim Bridger are that he was a great self-actional mountain man whose hunting prowess for stripping the West of fauna was of heroic proportion.   The truth of the matter is that Jim Bridger was a great transactional planner who organized several regional tribes and numerous European-Americans in order to obtain hides and furs for the European trading companies with whom he had legal contracts.   To fulfill the contracts, he made himself the chief across several tribes for the hunting and trapping.   The groups were tactically organized with specific geographical responsibilities.

Today, of course, hunting in the United States is predominately a consequence of government-sponsored activities with numerous governmental agencies coordinating activities with other private, quasi-governmental, and non-profit organizations.   Buffalo "hunting" is often conducted by the sponsoring agencies placing the "hunters" in the center of a pen with rifles and having the buffalo run around the perimeter until the hunters have killed their allotted number of buffalo. The amount of hunting and fishing depends on government funds allocated, the amount of game and fish stocked, and the numbers protected by the relevant agencies and organizations.   The third row of Table 3-1 is an abbreviated list of social institutions that provide for the hunting and fishing transactions.

Similarily, government plays a large role in the global financial sector. The banking system includes the Federal Reserve System, the International Group of Eight (G-8), the International Monetary Fund, the World Bank, the international electronic currency system, the local savings and loan (and therefore the Resolution Trust Fund), the Federal Deposit Insurance Corporation (FDIC), the Community Reinvestment Act, state governments, labor union pension funds, the bond market, U.S. Supreme Court decisions, property rights, and so forth. Field of concern, upon field of concern, upon field of concern; context over-lapped with context; criteria layered upon criteria—that is the transactional world of finance.

Further inquiry informs us that the lines between the areas in Table 3-1 need to be erased with regard to some concerns because education, hunting, and finance are all part of the same world. Federal Reserve policies influence economic growth, which influences family incomes, which influence activities for child development. Interest rates and bond rates help determine the cost of school facilities. Can the schools afford the new computers which help prepare children for the technological world of work? The schools educate those who become biological scientists, field workers, attorneys, conservationists, and managers for the wildlife agencies, as well as computer scientists, attorneys, economists, regulators, and managers for the financial sector. All three sectors function, on a daily basis, conscious of U.S. Supreme Court decisions. No simple self-action and interactional models can be found in reality. What do these examples teach us? They teach that we have, first, a need for the transactional approach of instrumental philosophy; second, that we need a problem orientation to know what to study; and third, that we need to seek complexity in our thinking about and modeling of reality. We need to "seek complexity and order it."[7]

The frustration of many, when the complexity of our ordinary world is presented during policy deliberations, leads those frustrated with the complexity of reality to invidiously admonish: "Follow the KISS principle. Keep It Simple, Stupid!" Of course, KISS is not a principle at all. It is contrary to scientific findings. It is usually the expressed frustration of someone who learns that society has found that the models used by the KISS proponent do not take into consideration either the complexity of reality or the consequences that society wants considered; for example, a frustrated engineer who wants a simple analysis while society wants ecological impacts, regional incomes, health, and community integrity also to be considered. Retorts to that KISS response that are equally invidious, but more appropriate acronyms, are KICK: "Keep It Complex, Knucklehead" and another KISS: "Keep It Safe Stupid." After the interactional balancing of acronyms, however, the difficult task of policy modeling still waits to be accomplished.

Although it is important to see the need to seek complexity and order it, the accomplishment of such a task is an expensive and difficult undertaking. To fulfill the need for modeling transactional complexity is the purpose of the social fabric matrix presented below in Chapter 6. The SFM is organized to provide the transactional approach with the analytical power to provide the needed transformative vision to solve problems.

Problem Orientation

Because of the great overlap among areas–as illustrated with the education, hunting, and finance examples–it is never possible to study everything connected to an area of concern. Policy inquiry would be impossible if all connections were to be pursued. Likewise, to recommend the study of an area, such as finance, is equally misleading because it is not possible to define finance except by defining a problem. For some finance problems, the bond market is of prime importance, while for other finance problems, social beliefs of rural people are the primary concern. The solution is to define the context of inquiry by the problem to be solved. The problem may be a theoretical problem or it may be an applied hazardous waste problem. In the policymaking process, "*problems* can be defined and depicted in many different ways, depending on the goals of the proponent of the particular depiction of the problem and the nature of the problem and the political debate. The process of defining problems and of selling a broad population on this definition is called *social construction*."[8] Without defining the problem, there is no indication of where to start, go, or stop in the policy research. "The social construction of a problem is linked to the existing social, political, and ideological structures at the time. The United States still values individual initiative and responsibility and therefore makes drinking and driving a matter of personal, not societal, responsibility."[9] Beliefs, interests, and desires guide the selection and definition of a problem. Thus, a housing finance problem will be defined very differently by a construction management group than by low-income housing advocates. A problem-solving orientation is neither radical nor conservative, but, rather, is aimed at "particular problems which arise on particular occasions."[10] Inquiry into a problem leads to an understanding of the consequences to expect from alternative solutions, and problem solving through knowledge creation leads to an accumulation of wisdom.

Consequences

Instrumentalist philosophy builds on the understanding that human social action has consequences. The purpose of policy analysis is to discover the consequences of particular actions, and to formulate policy so as to secure some consequences and avoid others. If we accept the fact that consequences are important, we can deal with problems and ask what needs to be done to obtain the consequences wanted or desired. The power to detect consequences varies with the instrumentalities of knowledge at hand. Therefore, to seek validity with regard to policy requires that we have the instruments to structure problems consistent with transactional systems. The purpose of this book is to provide for such an instrumentality.

The concern for consequences is a reason for the endorsement of democracy. Humans using democratic institutions will force the consideration of a broad range of consequences as well as the distribution of the consequences. Many approaches appeal to theoretical paradigms, or epistemology, or preferred ideology to justify policy. For example, some emphasize exclusive use of the market system as their model. This is inconsistent with instrumentalism and the concern for consequences. The market system is a particular institutional structure which ignores many consequences that society considers important. For this reason, democracy is essential to control the consequences of the market and to implement other institutional structures. Some problems can be solved—that is, consequences improved—with more competitive market activity, while other problems can only be solved with less market activity and more dependence on other institutions.

Those who demand a particular system or policy without regard to consequences are making decisions inconsistent with pragmatism and democracy.

Technology

Because of numerous attempts by scientists and philosophers, with both positive and negative connotations, to create a holistic philosophy from technology, a brief explanation is in order to clarify that such attempts are inconsistent with instrumentalism. Technology has been defined and conceptualized in numerous ways, sometimes confusing it with science, other times making it quasi-religious. One of the important conceptual breakthroughs contributed by Clarence Ayres was the understanding that technology is the *combination* of tools, skills, and knowledge.[11]   All three are present in all technology. Additionally, every new technology is connected to the past because new technology, as Ayres clarified, is a new combination of past technologies. Sometimes they are very different combinations, other times they are small changes in the proportions of current combinations. Potentially there is a wide array of permutations and combinations of tools, skills, and knowledge possible. This means that predictions about the direction of new combinations are impossible, because it is not possible to predict the kinds of combinations that persons in different situations in various institutions might complete. It also means that the technological discovery process is not subject to the law of diminishing returns, because the technological base continues to grow and diversify and, therefore, the potential for new combinations grows at a faster and faster rate. A historical continuum can be traced for all technology because new technology is a combination of prior technologies.

Technology, to use an old cliche, is the tail that wags the dog. Nothing has been more powerful in shaping historical change than the innovation of new technological combinations. The combination of the high keel of the Viking ships with the flat-bottomed ships of the Mediterranean changed the face of continents, because these ships had the capability to be sent out on the high seas. The combination of the steam engine and the spinning jenny led to numerous social changes, to include squalid living conditions during the industrial revolution. The "one-way" disc plow, that is, the combination of the rolling disc with the frame of the moldboard plow, played a major role in the creation of the "Dust Bowl" of the Great Plains of the United States. The series of combinations that gave us the machine gun made centuries-old war maneuvers obsolete. The combining of Boolean mathematics with the guitar transformed music forever, as the guitar was electrified.   The combination of the computer and telephone technology has created a twenty-four-hour-per-day gambling casino in international currencies, which has disconnected exchange rates from relative economic productivity.   If there is one issue that is clear, it is that technology when implemented is not neutral, although some have attempted to claim that it is.   New technology fundamentally and permanently alters human patterns of life and societal arrangements.   It is not neutral.

The neutrality argument, along with other philosophical claims, has been made because technology has become such a powerful and integral part of human lives. Four great minds of the Twentieth Century who dealt extensively with technology and its moral and philosophical meaning are Joseph Schumpeter, Clarence Ayres, Jacques Ellul, and E.F. Schumacher.

Schumpeter, who recognized how new technology destroyed institutions, made the evaluation that it was "creative" destruction.[12]   He arrived at this conclusion without the study of any particular technology or set of technologies and without establishing the philosophy, criteria, or political system that could have been utilized to arrive at such an evaluative judgment.

Ayres wrote at length about the destructiveness to the social and economic order caused by new technology. His writing exhibited full approval, and a bit of joy, as he explained the fall of one social order after another over the centuries due to new technological combinations. He offered a philosophical rationale for his delight.   Indeed, Ayres finally recommended that technological change itself be our philosophical base for determining what is morally appropriate.[13] Ayres' rationale was seriously flawed by changing the definition of the word "progress," by utilizing tautological arguments, and by equating technological change with instrumental value.

Ellul, quite in contrast with Ayres, found that technological change was morally vulgar. For him, the real problem was that modern society had made technology the center of all philosophy and social life. Modern society, therefore, was constantly devoted to the discovery and innovation of new technological organization to clean up the consequences resulting from technology's most recent societal dissolutions. For him, this drastically narrowed the human mind to "minding the machine" and arranging societal functions for the destruction of values, social institutions, moral standards, and so forth, as those entities interfered with new methods of organization needed for new technological innovations.[14]   Although a great deal can be learned from Ellul, he definitely overstates. We are surrounded with examples of societal control over and guidance of technology in order to protect and improve institutions, social beliefs, and the ecological system.

Schumacher correctly explained that technology, as such, was neither good nor bad, and that particular tool combinations needed to be evaluated for the social and ecological context in which they were to be used. For Schumacher, technology was to be guided, controlled, and regulated by humans through policy, and it was to be simply an ingredient in social and ecological matters, not the philosophical base.[15] Unfortunately, many of Schumacher's insights about technological policy were ignored because of the religious and ideological ideas in which his insights were wrapped.   He never carried his insights into a policy analysis paradigm.

Instead of trying to make technology and technological change into a mysterious force, evil phenomenon, philosophical base, or panacea for humankind, we need to understand that technology is one of the ingredients in the social matrix (see Figure 2-2).   Like other ingredients, technology needs to be researched and evaluated through a transactional analysis that is problem oriented and concerned with consequences. There is a need to assess the appropriateness of particular technologies that are being considered for adoption in a particular situation. Equally important is the need to recognize that new technological combinations are made by social institutions, that they are selected through judgments and decisions made within social institutions, and that they function and are maintained through the working of social institutions.   Society decides upon the technological combinations. Once those combinations are selected and adopted, requirements are placed upon social institutions.   Those requirements are usually not refined; they allow for latitude in institutional patterns. This does not mean that the knowledge of some potential combination that is not implemented makes for technology. Technology does not operate "on its own" as some self-actional or self-organizing force, and it does not function in some interactional relationship vis-a-vis society or the ecological system.   Although institutions are often changed to fit a different technology and thereby change society, the policy decisions for restructuring are made by social institutions. As Dewey emphasized, technology is directed and guided throughout the innovation and implementation process.[16] He argued that much of the direction for technology was based on inappropriate criteria, and he explained the need for the application of intelligence through instrumental evaluation. Whether or not an innovation should be considered good depends "on the direction which human beings deliberately give the change."[17]

Policy Criteria in a Transactional Context

Throughout the policymaking process, criteria need to be utilized. In spite of that fact, interest in the subject of criteria was scarce until the latter Twentieth Century. The works of Charles Peirce and Thorstein Veblen emphasized criteria—Peirce with explicit discussion of the character of criteria and Veblen with active application of criteria in his evaluation of various economies and institutions. Few scholars continued to emphasize their tradition. Thirty years ago, it was unique to find a discussion of criteria even briefly presented in books concerned with policymaking, planning, political science, economics, and the like. Today, such discussion has become much more robust. Given the fact that we are the political descendants of the Greeks, one might have expected evaluative criteria to have been a major concern all along.

As the interest in the subject has grown, so has the breadth of its definition. In current literature, the term "criteria" is often used interchangeably with standards, goals, decision rules, particle levels, and so forth. For the purpose here, the original definition of criteria as standards for judgment is recaptured. Policy evaluation is prior to and determines the establishment of policy goals, program standards, decision rules, and so forth. Or, stated differently, we need to judge and decide what policy is wanted before we can determine what goals, decision rules, or particular standards are to be implemented. For example, applying the decision rule of producing where marginal costs are equal to marginal benefits is not a policy judgment. The judgments have been made prior to the selection of that decision rule. Refined mathematical representations can be developed for the parameters and variables of some decision rules; policy judgments are not so rote or devoid of social process dynamics. United States Senator John Kerry recently articulated the difference well during a television appearance when, in response to a statement, he said, "I want to know how you made the judgment, not how you made the decision. What judgment and wisdom guided you?"

Normative Criteria Should Guide Research

We have learned from semiotics that a connection exists between the conditions of signification and the conditions of validity and verification. The interpretation of signs influences what is believed to be valid. The meaning of signs, objects, words, and ideas "is linked to a cultural order, which is the way in which society thinks, speaks and, while speaking, explains the 'purport' of its thought through other thoughts."[18] This means "every attempt to establish what the referent of a sign is forces us to define the referent in terms of an abstract entity which moreover is only a cultural convention."[19] The instrumental conception of criteria can be defined as an interpreter of social beliefs. Criteria stand between beliefs and interpretations of policy. Given the multidimensionality of beliefs and social systems, we need to take account of the multidimensionality of criteria. Normalized beliefs and social myths guide the selection of research problems and the articulation of normalized criteria. Normative criteria in turn should guide the research agenda if the research is going to be helpful in making policy for the relevant social context. Thus, we might visualize the process as displayed in Figure 3-1.

Criteria Are to Be Consistent with
Cultural Values and Social Beliefs

As outlined in Figure 3-1, there is a core of cultural values that are the ultimate criteria for the beliefs and myths of a social system. The anthropological literature, at least since the 1920s, has clarified the distinction between culture and society and therefore between cultural values and social beliefs. The cultural values are the dominant core and seldom change, especially in a policymaking period. Societies, as they change, through policy initiatives or otherwise, continue to be reformulated to make institutions and social beliefs consistent with the cultural values. Being aware of the cultural values and explicitly incorporating

Figure 3-1.   Social Policy A: Antitrust Area

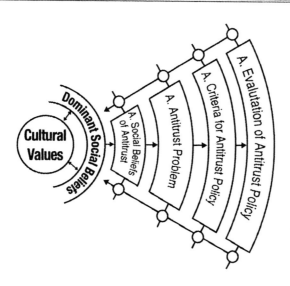

them into the design and determination of beliefs and policy criteria will assist in avoiding unnecessary social tension, alienation, strife, and numerous reformulations. The same is true for dominant core social beliefs if they are not going to be challenged or changed.

The traditional cultural core is expressed differently as society evolves, even though the cultural values remain stable. Closely associated with the core values are dominant social beliefs with which other social beliefs are to be consistent. In different areas, for example, antitrust policy or family policy, there are belief clusters that conform to the basic beliefs and values and are enforced in the institutional process for that area through legal codification, judicial decisions, working rules, and so forth. Figure 3-1 indicates a social belief cluster for the antitrust area. It is mainly due to infringements on the belief clusters that problems (meaning the failure of institutions, behavior, or attitudes to conform to beliefs) are defined.   Beliefs themselves can also be identified as a problem. Of course, there are cases where institutions and applied beliefs are inconsistent with the more dominant beliefs. Those are more serious problems (many times resulting in violent activity before the problem is solved) than when the inconsistency is with less dominant beliefs.

As displayed in Figure 3-1, the policy criteria for an area are to be relevant to the problem context and consistent with social belief criteria and with the social desires and ends that define the problem. Mark Okrent has explained that, for the instrumental pragmatist, the web of beliefs alone is not sufficient for consistency. "Rather, it is a web of beliefs and desires, or ends. Pragmatists understand mind and language in terms of action. When we think of behavior as action, we think of it as performed for the sake of some end, in light of some beliefs, and the behavior is action only to the extent that the beliefs and desires of the agent together make the action reasonable or rational."[20] Criteria are drawn in order to guide the action of evaluating to determine whether policy is reasonable. "In actual practice *the problem itself specifies or generates* (as inquiry proceeds) *the criteria of its resolution* . . . ."[21]   An example is the ongoing evolution of criteria creation and evaluation for policies to develop apples.

Policy criteria for apples have not been designed by looking at apples, nor are they designed by a self-actional agent, nor an interactional balancing of supply and demand. First, the criteria for a good apple in the United States are the consequence of many different overlapping institutions, knowledge bases, and competing interests that include government agencies, research universities, regulatory agencies, consumer protection groups, health advocates, technological systems, labor unions, nutrition researchers, growers' cooperatives and their attorneys, environmental protection agencies and their attorneys, foreign trade inspectors, advertising agencies, fertilizer and pesticide production corporations, and so forth. And there are lobbyists, scientists, government and university analysts, with billions of dollars associated with those institutional processes. They are all involved in designing, refining, testing, creating, and applying the criteria for judging apples. Second, the apple design criteria that are conveyed to scientists in universities, such as shelf life, color, shine, susceptibility to bruising, nutrition, size conformity, herbicide and pesticide tolerance, and health risk, do not come from the apple. We do not observe apples to find the criteria. The criteria are determined through socioeconomic processes. We did not need to find an apple to know a good apple; the kind of apple wanted did not exist. It was created by policy-driven science.

Advanced technology, of course, plays a major role in our beliefs about the kind of apple tree that ought to exist. When it became clear that a machine could grasp a tree by the trunk and shake the fruit from it, the question became, can a tree be designed to withstand the shaking and its fruit to resist the bruising? The answer in some cases has been yes.

Consistency between criteria and policy evaluation is not circular. It is not circular to say that a tractor and a plow must be technologically consistent to function together. A call for linkage does not mean that a tractor and plow are the same. Likewise, it is not circular to say that we can judge policy as acceptable when it is consistent with criteria. Criteria stand between the social system and the policy evaluation. "Instead of there being anything strange or paradoxical in the existence of situations in which means are constituents of the very end-objects they have helped to bring into existence, such situations occur whenever behavior succeeds in intelligent projection of ends-in-view that direct activity to resolution of the antecedent."[22]

Social beliefs are given by society and serve as the criteria for policy criteria. Figure 3-2 (of which Figure 3-1 is a part) clarifies that the policy criteria for evaluating antitrust policy is to be consistent with the beliefs, policy criteria, and evaluation in all the other societal areas in Figure 3-2. For example, antitrust policies could allow prices to reach such a level that most families could not maintain a real income sufficient to support family needs. An example is the failure to enforce antitrust regulations of prescription drugs, thus allowing prices in the United States to rise to a level from six to ten times greater than those in England and Germany, respectively, for the same drugs. Or, on the other hand, policies enforcing fierce competition may push prices so low that corporations will destroy the ecological system (Social Policy E in Figure 3-2) in order to lower costs to meet the low prices.

Another aspect with which criteria are to be consistent is with different social arrangements and different social beliefs of different groups that share the same cultural values. The multidirectional concerns in establishing criteria are not limited to cases where there are common beliefs. This could be demonstrated by the overlap of more than one Figure 3-2, each with the same culture but different societal beliefs. The concern for pluralism "entails the use of multiple *sets* of evaluative criteria."[23] for resolving policy questions. Within a nation with a common culture, different beliefs and institutions exist among different groups if freedom of varying associations is allowed, for example, distinct religious groups that have the same culture. Institutions and laws can vary from nation to nation with the same culture, or from

Figure 3-2. Six Integrated Social Policy Areas

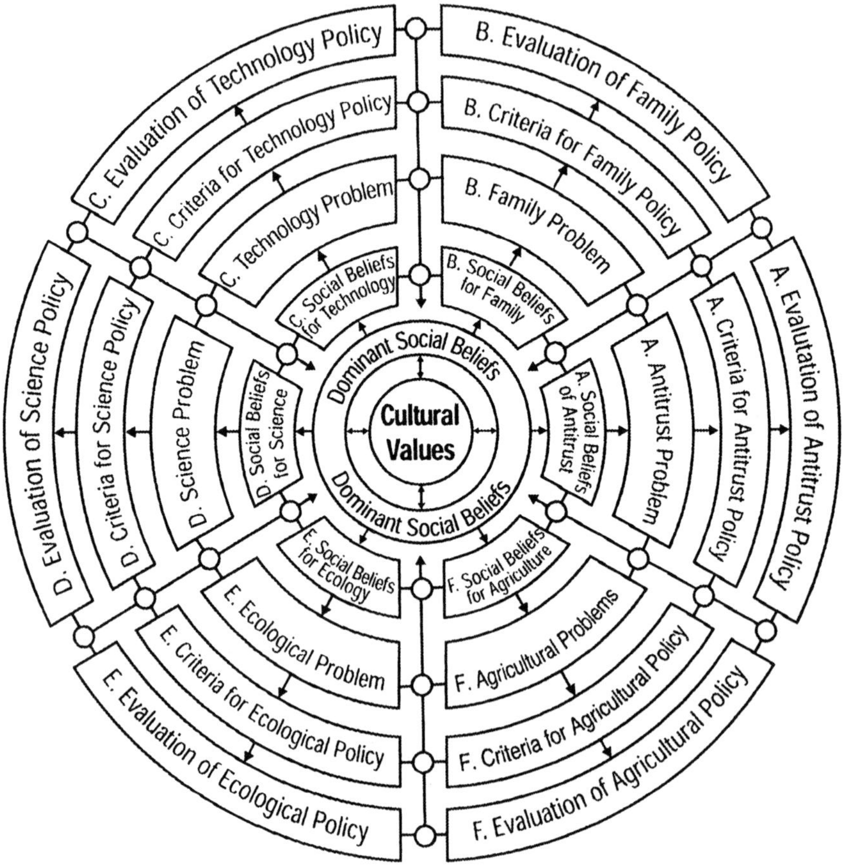

state to state, region to region, or religion to religion with the same culture. In the United States, agricultural policy that fits the Amish of
Pennsylvania may not fit the wheat farmers of Montana or Cargill's
corporate hog factories of the High Plains. If free association and differences in the structure of association is a Western belief, then policies
should reflect those differences, and the formation of policy criteria for
evaluating them should be guided by the differences. As the technological society advances, we can expect even greater variation in the
kinds of association and beliefs among groups.

   Finally, formation of policy criteria is constrained in some areas
by the overlap of policies with other cultures, either within a nation or
transnationally. This multicultural concern could be demonstrated with

the overlap of more than one Figure 3-2, each with different cultures and usually with different societal beliefs. Within the United States, the Amish farmer in Pennsylvania and Cargill's corporate pig factory executives are members of the same Western culture. Yet within the United States, cultures different from the mainstream culture with different cultural values are a reality; for example, the nations-within-the nation of the descendants of the original tribes that existed prior to European immigration. Likewise, cultural differences exist between nations, as in the case of India and the United States, and those differences are crucial when formulating global policies. In summary, differences in cultures and societies, and the institutional overlap of the same, requires that policymaking be an arduous task. The arduous task is to define, design, and apply criteria consistent with all the overlapping policy concerns for each particular social group across various groups and cultures. The issue is one of constraint, and whether a set of criteria is adequate depends on what surrounds the policy area. If the real institutional constraints and belief criteria are not recognized and heeded, some groups will be severely harmed as policy criteria are narrowly focused on misguided beliefs.

No better example of narrowly focused and misguided criteria can be found than in the well-known memo of Lawrence Summers featured in *The Economist*. In the memo, Summers indicates that he selected money-making as the dominant policy criterion and, therefore, judged that it was efficient to create health hazards for low-income people in Africa and ecological damage for their habitat.[24] These kinds of conclusions are not new in neoclassical economics. The same policy criterion of money-making was utilized by Burton Weisbrod a few decades prior to Summers, to demonstrate that the pecuniary value of keeping young white males in school was four times that for young black females. Thus, he concluded that no special effort should be made to educate the latter.[25] The lack of concern at the policy level for multisocietalism and multiculturalism results from a lack of emphasis on multidimensional criteria in policy evaluation.

One of the reasons that narrow and misguided criteria, abstract metaphors, and simple models have developed that are so misleading for policymaking is because analysts have divorced themselves from the institutional reality of policymaking. The more that policy analysts are in the real world bumping shoulders with labor leaders, observing street gangs, being handled by lobbyists, haranguing policy assistants to find a database, meeting with religious leaders, calling think tanks and university professors for relevant theory, discussing problems at steel mills, arguing with ranchers at horse shows, being heckled at commission meetings, being challenged at legislative hearings by politicians, visited by advocacy group attorneys with threats of lawsuits, and facing opposing scientists on television—the more there is of real-world involvement—the less there is an opportunity for simplistic criteria, metaphors, and models to take over. Charles Anderson referred to the problem of the policy scientists divorcing themselves from reality as "the problem of overpractice"[26] in which the policy scientists are so intent on rational perfection of technique that they build esoteric systems that are insufficiently cognizant of social reality.

The more abstract the policy models, the less reliable; they often do not even give the correct direction. Abstraction means "removing the clutter." That clutter, however, is the social life process, and no idea has legitimacy except as it is embedded in a life process. Power bumps, mental jostling, and legal body checks on our body of theory all help to dissolve simple abstract metaphors and replace them with complex ones corresponding to the real world. If policy criteria are to be based on social ideas, social experience is necessary for designing them. As Charles Peirce emphasized, ideas are a semiotic product. For policy criteria, ideas need to be the semiotic product of a heavy dose of real-world experience.

Criteria and Contextual Shift

If modeling that is too abstract or esoteric can be avoided, there is still the intimidating task of designing criteria consistent with the new beliefs and institutions that will be necessary to solve a problem. Making new beliefs the locus of concern can be referred to as the "contextual-shift" aspect of policy criteria. Earlier it was explained that criteria are to be consistent with belief clusters in the relevant context, yet problem solving usually requires changes in institutions, beliefs, and technology—that is, it requires changes in the context. There is no contextual shift if the problem can be solved by selecting policies that will align and strengthen the bonds of current institutions and beliefs. If, however, the socioeconomic problem is more pronounced, it is usually a mistake for the policy criteria to be made consistent with *current* beliefs. To utilize such criteria to select policy arrangements will not solve the problem. Institutions usually need to be changed to solve problems. This means a major task of any policy analysis is to design a set of criteria that will be consistent with the *new* set of beliefs and institutions necessary for solving the problem. As the contextual shift is taking place, new technology or other disruptions can change the context markedly. This is why technology assessment, prior to innovation, is crucial. The faster disruptions happen, the less opportunity for the instrumental research process to gain an understanding of the most reasonable policy and the less opportunity for the policy to be successful.

The contextual-shift approach to policy analysis calls for an extensive and expensive investment in policy research. Although expensive, the alternative of committing billions of dollars on resources for misaligned policies that fail is even more expensive, very frustrating, and harmful to citizens. The research is necessary for designing and refining the criteria so that policy will lead to a new context consistent with the needs of the whole social matrix. Irrespective of the extent of research, an overriding pragmatic criterion is that "it is not until after one has acted on certain hypotheses that the situation will eventuate which will either confirm or disconfirm the proposed solution."[27]

One of the basic assumptions of democracy is that we will make policy mistakes. That is one of the reasons the quick reaction allowed in a democracy is necessary. Given the task at hand, mistakes will be made because the policy criteria were designed for a world that does not yet exist. When we begin to implement that world, there are sure to be mistakes in both the criteria that were selected and the institutional arrangements. To fulfill the iterative process, procedures for constant monitoring of policy consequences and subsequent reformulation of policies and programs are necessary. "One cannot know prior to taking actions whether they will eventuate in the desired consequences because the confirming state of affairs not only does not exist when the question of the truth of a particular solution is raised, but *its very existence* is contingent upon many interrelated factors, one of which is the particular procedure or procedures one takes . . . in dealing with the problem."[28] The function of democracy cannot be fulfilled in a vacuum. Policies must be implemented to be tested and to improve social modeling. If social modeling is not ongoing, guidance will not be available for the democratic policymakers. This means a major part of policy analysis will be to study the community structural changes that take place as a result of the application of a set of criteria and their resulting policies, and to change the criteria if they lead to undesirable consequences.

Criteria as Process

No attempt is being made to contend, as some who have tried to save the foundationalist view have stated, that criteria can be interpreted and applied objectively. First, objectivity is not desired; the purpose is for normative criteria to be applied. (In addition, no superior group with the powers of objective interpretation and application is being assumed. Persons with the kind of personality or ideology that leads them to want to pursue objective truth should probably be considered dangerous because too often they think they find it.) Consistency with social criteria is the goal of policy judgments. The refinement of the interpretation and application of criteria is achieved through discretionary processes made up of legislative bodies, judicial proceedings, research inquiries, advocacy efforts, and so forth. The validity of the interpretation and application of criteria depends on the extent of social processing, or social interference if you will. Marc Tool explained that what is required is a criterion of judgment that draws on and is reflective of experience and a continuously refined product of reasoned reflection.[29] The reflection of experience and the resulting challenge to criteria and their interpretation come swiftly in a democracy, as do the counter challenges. The challenges are swift because consequences are real and not necessarily pleasant for all parties involved. The process, called democracy, is what some refer to as being too "messy" to allow for a policy science. Quite the opposite is the case. The democratic process refines criteria so that more refined analysis and evaluation are possible. In the policy world, refinement and order are the consequences of conflict within discretionary processes.

The One Best Way Is Not Viable

Conflicting strands have been interwoven through the social sciences that revolve around whether there is "one best way." One strand has emphasized that new technology requires a new social structure and process that is consistent with the new technology. This often implies that there is one best way for society to be structured. Coexisting with this is the view that instrumentalism requires democracy, which allows for alternative kinds of associations. In a democracy, different groups structure their associations and behavior quite differently. The Tavistock sociotechnical studies in Europe found that automobile factories with similar technology in different countries have very different social structures and social relationships in the factories. In addition, from observation we recognize that modern technology encourages even greater diversity among the social structures of different groups. Much of the frustration of Western global policy analysts, after the dissolution of the Soviet-United States dominated system, exists because so many countries and social groups have structured themselves according to models that do not coincide with the "one best way" envisioned by Western analysts. The general problem with much analysis is the "search for criteria that are appropriate to all cases in the genre."[30] As Sadler explains, it is a mistake to assume that "commonality" is the principle that is taken to validate a set of criteria. This is not possible unless the cases are the same. "But as the variety of cases expands, the probability that a common set of criteria will be uniformly applicable decreases, simply because the number of potential criteria which could be drawn upon is large."[31]

It may be that the idea for the "one best way," or "the optimum" as the neoclassical economist would say, comes to us not from observation, but rather from the same tradition that gave us "objective truth." "Beginning with Plato, the Greeks fashioned their gods according to mathematical form, and the Romans and Christians continued the tradition. This was carried into science and analysis in the Western world because early science was an extension of religion, and like religion, science was searching for the truth—the one truth—the one objective truth according to mathematical form."[32] The mistaken idea that there is one acceptable set of beliefs and criteria has had a long history in philosophy. Chisholm stated this mistaken idea as follows: "If we find a pair of beliefs that *contradict* each other, we will reject at least one of them."[33]

If we believe in (or have) discretionary democracy and free association, and if new technology allows for more diversity, then numerous different social structures are viable. Thus, the job of designing policy criteria also becomes much more difficult. A single set of criteria for health care policy, for example, is not possible. A different set of criteria is needed for each context, and a still different set where the contexts overlap. What will be attempted to be accomplished with policies in the United States, for example, will differ depending on whether it is for the Amish, the Mormons, recent Somalian immigrants, the Winnebago-Americans, or Orthodox Jews. It may be satisfying to read neat formulae as often presented in policy science literature, however, the formulae accomplish little more than to please psychological predilections. Reality is much more complex and detailed. That detail is what planners and civil servants work with in the bureaucracies of our city, county, state, and national governments, where they are constantly monitored and challenged on a daily basis by the citizenry and the citizenry's covey of lawyers.   The reality of policy criteria is the detail of numerous different kinds of group contexts.   The only alternative for avoiding this kind of messiness (as it is often called) is the "one-best-way" solution, as has been attempted during some of the darker and more brutal eras of human history.

For too long, the general idea that compromise and common ground are the solution has frustrated good policymaking. While compromise and common ground are essential for a democracy to function, that does not mean that one policy or a unified common ground with regard to normative criteria should be sought and enforced throughout the social setting. Former President Jimmy Carter once stated with regard to diversity of beliefs, when interviewed on a televised news report: "Diversity need not be a handicap." Then he added, "We can find a common ground." No.   The two statements are in conflict. That they are in conflict logically is not the concern. The important concern is that the statements are in conflict with designing relevant criteria. This kind of "common ground" thinking leads either to abstract, mushy criteria that will not fit any context well, thus accomplishing very little, or to a firm set of criteria that ignore the beliefs and unique associations of certain groups. When such groups are silenced institutionally by being ignored, criteria cannot be properly drawn.   Such marginalization, in addition, leads to anger and civil unrest, a situation that has become more common. The ripping and tearing of our social fabric because of cavalier criteria contribute to the social disruption and violence that can engulf local communities as well as global relationships.

Mandates

Criteria are also relevant to the arguments about mandates in a democracy. Difficulty occurs when mandates from the central government are stated in terms of rules, flow levels, and requirements to take a particular action or to establish a particular technical procedure; for example, when the central government requires that every city of a certain size should construct a waste treatment plant of a particular kind or that every student attending school should be vaccinated. Such mandates ignore contextual variety. Treatment plants may not be necessary if the city does not generate a high level of waste or if the natural environment is robust enough to cleanse the water without treatment. Every student may not need to be vaccinated if some are from families that have already provided for their vaccinations.

The more instrumental approach in a democracy is for the central government to define and establish criteria for local governments, corporations, and families to use in making judgments and decisions; for example, criteria about being free from a particular disease or maintaining clean water. In this way, local institutions and decision makers can be more creative in designing alternative ways to meet the criteria. More importantly, they can respond to the criteria consistent with the relevant context.

Because of the focus on inquiry and validation in instrumentalism and because of the community orientation recommended for policy analysis, there are numerous safeguards within the process itself against the subjectivity of an individual local government gaining the upper hand. Schlagel has stated that because problems are identified by the community in a democracy, and the criteria of their solution are subject to external review with publicly accessible evidence, the application of applied pragmatic criteria is no more susceptible to subjective abuse and is no less rigorous than that of any other criteria.[34]

In contrast to the centralized mandate approach, some champions of local government have overlooked the necessity for the central government and constitutional bodies even to mandate criteria. An example is Eleanor Ostrom, who, in her *Crafting Institutions*, a book that is generally helpful, explains that in crafting institutions for irrigation systems, all the multiple layers of government should be making rules for the success of such irrigation systems. She outlines three levels of rules: (1) operational rules, which directly affect the day-to-day decisions of water users and suppliers at the local level; (2) collective-choice rules, which indirectly affect operational rules through policy-making, management, and policy adjudication rules; and (3) constitutional-choice rules, which determine who is eligible to partici-pate in the system and what specific rules will be used to craft the set of collective-choice rules, which in turn affect the set of operational rules.[35]

Ostrom's paradigm depends exclusively on rule making, an excessive dependence for two reasons. First, constitutional action is finalized mainly through judicial judgments, and the latter are mainly finalized by the provision of criteria for other bodies to use in making decisions. Thus, the idea of constitutional rules being the dominant mode is inconsistent with the experience mode of judicial bodies. Second, Ostrom has indicated that most rules must be made at the local level in order to allow for the diverse kinds of decisions that must accompany the diverse contexts at the local level. As she correctly states, rules codified by external administrative agencies, national legislation, and the judicial arena rarely reflect the particular circumstance of a particular system.[36] If most rules must be made at the operational level, and if mandates from the collective legislative and judicial bodies are needed, then how can the desired effect be achieved when bodies have been provided only with rule-making functions? It cannot. Thus, the collective policymaking and judicial bodies need to depend more on criteria. Criteria can be established without establishing special rules; thus, creativity can be used to define rules, technology, and institutions that are consistent with the criteria and with local situations.

The neoclassical tradition in economics fails to recognize social system criteria and norms and instead concentrates on rules—rules justified in terms of pecuniary enhancement without reference to social morality. For neoclassicalists, the rules are to guide procedures without concern for social or ecological outcomes or consequences. Of course, without criterial norms, it is not possible to make judgments about what rules ought to exist and about whether the consequences from the rules are consistent with the normative components. Market rules are emphasized in the neoclassical paradigm so as not to have to judge market consequences by society's moral and ethical criteria established to judge the legitimacy of markets. Conclusions are not possible when analysis is separate from either the relevant context or consequences. To separate the analysis of rules from their real-world context and from the impact of those rules on society leads to degradation of society and limits the possibility of salutary policymaking.

Science Is a Policy Area

Since science plays a major role in modern policymaking, it is important to recognize that there is considerable pluralism in science with regard to the set of social beliefs and resulting criteria. Science is an area of policymaking, and scientists aspire to be accurate policymakers with explicit criteria. A spectrum of writing, ranging from that of Gjessling Gustrom to that of Stephen Jay Gould, emphasizes that science is influenced by the conceptual system of the culture and society. Science, like other policymaking areas, is directed, constrained, and controlled by normative social criteria. It is conducted by a community of inquirers with similar policies, strategies, and tactics for scientific activities.   "The belief in an independent, self-subsistent universe, knowable at least in principle, provided the ontological setting for most scientific inquiry and philosophic speculation in the West, until recently, as Dewey persuasively argues in the *Quest for Certainty*."[37]   A few decades ago, young recruits in economics were drilled in the beliefs of the objective logical positivists with a catechism of belief criteria about coherence, correspondence, external verification, a specified logic, and so forth. Ernan McMullin said of logical positivism: "It was, perhaps, the most ambitious foundationalism in the entire history of philosophy, outdoing even that of Aristotle. And as we know, it collapsed."[38]   Science today is organized and directed around an array of various sets of normative propositions that, in turn, are organized around an array of different integrated sets of beliefs, and is practiced through an array of different contextual criteria. "There are many other criteria involved in science besides those of valid argument."[39]   Instrumentalist science, as a world theory, is holistic, integrative, and evolutionary. "Insofar as a world theory stakes out intellectual claims, either explicitly or implicitly, it purports to be knowledge, and as such it is subject to the criteria for meaning and knowledge whatever those criteria may be."[40]   Scientific criteria are policy standards for scientific rationality and interpretation. "The mistake of the logical positivists was to reduce rationality to logicality in the hope of making verification a simple and non-controversial affair, thus making possible a conveniently sharp line of demarcation between science and the fuzzier sorts of human activity. But even at the very level of observation, there is the matter of *choosing* the concepts in terms of which the observation will be expressed . . . ."[41]

Because of the relationship between science and the contextual framework, "science cannot, in consequence, be constructed in logicalist or foundationalist terms."[42]   This means we should no longer talk in terms of truth or not truth, but, rather, in terms of being valid or not valid, or making warranted or nonwarranted assertions. Are the findings valid or warranted in terms of the normative criteria and context selected? Foundationalists, like their fundamentalist counterparts in the religious world, claim there are basic foundations or fundamental truths that are given. When we observe more closely, however, what is assumed to be truth apart from societal beliefs is usually consistent with dominant societal beliefs and cultural values.

As with other policymaking areas, human institutions provide the criteria and context of science. Scientific criteria and context are policy decisions made by the bureaucracies of research universities, corporate financing sources, and other institutions. Today, the normative criteria of science are dispensed more and more, along with the massive sums of money for doing the research, from government bureaucracies that have the legal and ethical responsibility to direct scientific research. In addition, scientific journals endorse policy criteria directives through the articles accepted and through the evaluation process.   Science does not take place in a vacuum.   The actors called scientists are cultural and social actors. Science is organized in social institutions; sciencing is a social policy process. Thus, the fierce intellectual battles about policy science criteria are important.

The basic social beliefs and normative policy criteria for conducting scientific inquiry are extremely important today because scientific findings are very influential in general, and because different scientific criteria give us different social results. If we return to Figure 3-2, we are reminded that scientific policy criteria inconsistent with other areas will likely damage the other areas. For example, late in the 1800s and early 1900s, scientists generated I.Q. test scores and related experimental results from which the scientists claimed to have found some ethnic groups to be mentally inferior.   Had that evidence been allowed to stand, the United States could not have survived as a democracy.   Likewise, our system will not be allowed to prosper if the neoclassical standard of the market is to be the scientific criterion for making judgments about ecological systems policy, educational policy, international trade policy, or the like. Kathryn George emphasizes that what is defined to be sustainable agriculture depends on which members of the community select the criteria to guide the scientific work for determining sustainable practices.[43] The same is true of other areas. Scientific knowledge is not discovered; it is created through the application of beliefs and scientific policy criteria. The models that are identified as the context corresponding to the relevant problem are created and selected by the scientists. This means science, and therefore scientific knowledge, is not a consequence of idle curiosity. "Science advances by adopting the instruments and doings of directed practice, and the knowledge thus gained becomes a means of the development of arts which bring nature still further into actual and potential service of human purposes and valuations."[44] Thus, science is a directed activity.

Conclusion

For the directed activity of policymaking to be successful, criteria are needed for judging the various alternatives available to achieve a solution to the problem at hand in a manner that is consistent with the communities' standards of morality and justice. As stated above, for policy criteria to fulfill this role, they need to be embedded in the context of pluralistic social beliefs and overlapping institutions.

The approach explained here is inconsistent with the narrow ideological approach so prevalent in neoclassical economics in recent decades. Neoclassicalists have endorsed a narrow set of policy rules to confine and overpower the rich criterial texture found in our pluralistic world. Karl Polanyi, in his *Great Transformation* (one of the most important books written in the Twentieth Century), explained the dangers of allowing the narrow rules of a market system to overpower social criteria and social rules. To avoid the repression and subjugation of the community requires that the multitude of social beliefs be given standing in policy judgments. This is an arduous policy task, but the alternative is the destructive approach of adopting criteria external to contextual reality. Formulating criteria for an instrumentalist policy framework is a huge research undertaking; however, it is exceeded by the larger adverse policy consequences that follow without it.

Concurrent with the development of instrumental evaluation since the late 1800s has been the development of general systems analysis (GSA). Although researchers and advocates in both areas were inspired by the recognition of transactional processes and holistic systems, there has been little intellectual cross-fertilization between the two endeavors and few attempts to integrate their knowledge bases for policy analysis. Some of the general systems principles are consistent with instrumentalism and are presented next because they can assist as organizing theories for the application of instrumental philosophy.

CHAPTER 4

GENERAL SYSTEMS PRINCIPLES
FOR POLICY ANALYSIS

General systems analysis (GSA) is based on principles that are relevant to all systems whether social, biological, technological, ecological, or economic. The GSA principles presented here are a set with which methodologies and models need to be consistent if those methodologies and models are to be useful in explaining and evaluating socioecological systems.

Through the years, knowledge, scientific methodology, and experience accumulated across scientific disciplines and converged to find commonality among systems. Common systems principles and characteristics, such as openness, complexity, wholeness, hierarchy, and regulation were found to be useful in explaining the functioning of all systems. In 1941, Andras Argyal stated that "with regard to dynamic wholes, one would expect that a given part *functions differently* depending on the whole to which it belongs. We would expect that the whole has its own characteristic dynamics."[1] The commonality of these dynamics, according to Daniel Katz and Robert L. Kahn, allows for them to be described as systems theory. "Systems theory is basically concerned with problems of relationships, of structure, and of interdependency rather than with the constant attributes of objects."[2]

The function of GSA in evaluating government programs, social costs, public goods, and environmental policy is as a tool kit of principles for understanding systems. The principles are to be used to describe and explain the working of socioecological systems in order to allow for the evaluation of the system and its parts, and allow for a system to be compared to alternatives. The principles are not just a descriptive nomenclature. They are theories for organizing analysis, explaining systems, and judging policies. To assist in understanding the relevance of GSA to the evaluation of systems, 12 relevant systems principles and their characteristics are defined and explained here.

System Defined

✗"A system is a set of objects together with relationships between the objects and between their attributes."[3] Objects are the elements and components of the system. Attributes are the properties of the elements and components, and relationships are what tie the system together. The relationships to be considered "depend on the problem at hand, important or interesting relationships being included, trivial or uninteresting relationships excluded."[4] To use Kenyon De Greene's definition, "in the most general sense, a system can be thought of as being a number or set of constituents or elements in active organized interaction as a bounded entity, such as to achieve a common whole or purpose which transcends that of the constituents in isolation."[5] There is no end to a system. Any relationship or delivery between components leads to additional deliveries, and to positive and negative feedback deliveries. One-dimensional systems (such as would be implicit in an assumption that human consumption is the end of economic activity) are not real-world systems.

Openness

All real-world systems are open systems, and all open systems are non-equilibrium systems. "Open systems are those with a continuous flow of energy, information or materials from environment to system and return."[6] There are misconceptions which arise, both in theory and practice, when social organizations are regarded as closed rather than open. "The major misconception is the failure to recognize fully that the organization is continually dependent upon inputs from the environment and that the inflow of materials and human energy is not a constant."[7] Systems and their environments are open to each other, and subsystems within the systems are open to each other as well. Living systems both adapt to their environment and modify their environment.

GSA divides the analysis between the system under consideration and the system's environment. The system description is referred to as the internal description, or the *state* of the system. However, all systems are influenced by an *external* description, which is outside the boundaries of the system. For example, a wetland ecology receives inputs such as contaminants, sediment, and nutrients from agriculture. Although inputs from (often called forcings) and outputs to (often called responses) the external environment are important to the system, no attempt is made to model the systemic structure of the environment itself.   The environment has only a functional "black box" purpose to the system. The term "environment" as used in systems analysis may mean an ecological system if, for example, the system under study is a socioeconomic system.   If, however, the system of interest is an ecological system, then the socioeconomic system is the environment.

In systems analysis, environment refers to the functional area outside the system. Because real-world systems are constantly open to their environments, systems cannot reach an equilibrium state.   One of the goals of analysis is to be able to match up the two kinds of system descriptions. "The external description is a functional one; it tells us what the system does, but not in general how it does it. The internal description, on the other hand, is a structural one; it tells us how the system does what it does . . . ."[8]

Four external functions of the natural environment for the social system have been explained by James A. Swaney.[9]   The functions are: *Natural goods production*, which includes wilderness, greenery, landscape, scenery, and so forth.   It is often competitive with natural resource utilization, and is restricted in quality and quantity by the production of effluents from households and production centers.

Natural resources, which includes the raw material and energy sources taken from the ecosystem, upon which the production of goods and services is dependent.   Natural resources represent only part of one of the two flows from the environment to the economy. They flow to the private and public production centers.

Living systems or life support services, which are the services necessary for life in the environment, human communities, and work places. They include oxygen for workers in the economy and carbon dioxide that is "breathed" by farmers' fields. Life support services provided by the environment are hampered by growth in the production of economic goods.   The "key point is that the life support system *cannot* be . . . priced or otherwise allocated by the economy." [10]

Sink function, which refers to the fact that all wastes from all parts of the environment and from the economy are disposed of in the environment. This sink function can no longer be taken for granted, because overloading the sinks with wastes and pollution from the households and production centers increasingly interferes with the environment's other three functions. Among the most important natural resources used up have been the living systems destroyed by industrial waste disposal.

Nonisomorphic

Real world systems are not isomorphic from part to whole. Isomorphic systems are systems in which the whole is a reflection of the parts; for example, the sum of the parts. The idea that systems can be studied by looking at individual parts is often referred to as reductionism. The nonisomorphic, or holistic, approach to analysis has been viewed by scientists and policy analysts as a major departure from earlier mechanistic and reductionist thinking. In living systems, the parts work according to the structure of the system. Work procedures are guided by the requirements of technology, and human consumption is guided by social requirements. GSA allows investigators to accomplish two procedures very important to an investigation. First, it allows for identifying and defining the system of the problem area, which is embedded in the overwhelming complexity of the real world. Second, it provides a means of disaggregating the system into subsystems without practicing reductionism. As Robert Rosen explained, a reductionist hypothesis cannot be true for many of the defined properties of greatest interest about systems.[11] The task, thus, is to disagggregate or fractionate a system into nonisomorphic systems so that "(a) each of the fractions, in isolation, is capable of being completely understood, and most important, that (b) *any* property of the original system can be reconstructed from the relevant properties of the fractional subsystems . . . ."[12] In this way, subsystem systems can be effectively analyzed consistent with the original system.

Equifinality

The equifinality property of systems means that open systems "can reach the same final state from differing initial conditions and along a variety of paths . . . ."[13] Because systems are not automatic equilibrium systems, they respond to changes in the external environment to achieve a system goal. Only by adjusting the system can open systems attain a steady state. The degree of equifinality is reduced as more control mechanisms are introduced.[14] For example, if a technology rigidly sets the requirements of the social system, the flexibility of the social system in dealing with pollution is reduced.

The concept of equifinality becomes important, for example, when determining the restoration of an ecosystem. Since there are alternate paths to achieving system viability, some paths may be less expensive to restore in terms of resources than other paths.

System Components

Real-world systems studies, whether they are called sociotechnical, socioecological, or socioeconomic, are concerned with the integration of the components of the social, technical, and natural ecological subsystems. The components of systems, as stated in earlier chapters, are: (1) cultural values, (2) social beliefs, (3) personal attitudes, (4) technology, (5) social institutions, and (6) the natural environment.

Control and Regulation

Crucial to systems, and therefore an important focus of GSA, is the control and regulation mechanism of systems. System control and regulation takes place through rules, requirements, and criteria. Two types of controls are emphasized in GSA.

The first type of control is that every system element or subsystem which makes a delivery to another element or system exerts control "if its behavior is either necessary or sufficient for subsequent behavior of the other element or system (or itself), and the subsequent behavior is necessary or sufficient for the attainment of one or more of its goals."[15] This is control through relationship and requirement linkages. An example is the effect of habitat cover on the kind and structure of wildlife in a habitat. Different kinds of ground cover control for different kinds of species populations. Before elements or systems can perform the behavior pattern, which gives them the first type of linkage control, however, other control mechanisms and rules are needed to determine their behavior. These constitute the second type of control. "Biological and social structures are not objective in the sense of physical laws. They are coherent systems obeying dynamical laws and syntactical rules that are distinguished from isolated physical systems by their ability to change their internal constraints and thereby change the rules of the game."[16]

The functioning of DNA is an example of system rules which give DNA extraordinary authority over cellular collectivity, and "the development of multicellular organisms . . . shows that the cells do not simply aggregate to form the individual, as atoms aggregate to form crystals. There are chemical messages from the collections of cells that constrain the detailed genetic expression of individual cells that make up the collection. Although each cell began as an autonomous 'typical' unit with its own rules of replication and growth, in the collection each cell finds additional selective rules imposed on it by the collection which causes the differentiation."[17]   The presence of controls and constraints in a system is a distinguishing characteristic of living systems.

Technology is another example of system control. It provides requirements for social systems. These are often in the form of criteria, which must be met (see Figure 2-2). The technical component "contributes preeminently to the self-regulating features of the system."[18]   In this way, "the technological system sets requirements on its social system and the effectiveness of total production will depend on how adequately the social system copes with these requirements."[19]

In social systems, primary controls are social belief criteria. They give the social system structure. Social structure includes constraints, rules, customs, beliefs, legal codes, and the like. These structure social systems by guiding social and economic action, by legitimizing transactions, and by requiring deliveries to be made. As clarified above, in addition to the cultural, technological, and social, there are ecological system constraints. These are also part of the system, and when overridden, the system is degraded.

Hierarchy

In view of system control, it is probably not surprising that all systems experience hierarchical arrangements of many kinds. Laszlo defined hierarchies as "higher order systems, which within their particular environments constitute systems of still more indecisive order."[20] Howard Pattee, as well, emphasized the control of hierarchy in systems.

"In a control hierarchy the upper level exerts a specific dynamic constraint on the details of the motion at a lower level, so that the fast dynamics of the lower level cannot simply be averaged out. The collection of subunits that forms the upper level in a structural hierarchy now also acts as a constraint on the motions of selected individual subunits. This amounts to a feedback path between levels. Therefore, the physical behavior of a control hierarchy must take into account at least two levels at a time."[21] The feedback path among levels needs to be taken into consideration when modeling systems for policy analysis.

Flows, Deliveries, and Sequences

Systems might be identified as flows of sequenced deliveries. The concept of flow is fundamental to systems. "Internal and external descriptions of systems are wholly complementary approaches to modeling systems structures and this equivalence can be seen through the unifying concept of flow. If a system has been described internally in terms of a number of state variables between which are defined certain relational functions, then these state variables can be considered to change as results of flows occurring."[22] It is important to include input flows from the natural system delivered to socioeconomic systems in order to complete analysis of a socioeconomic system. Likewise, it is important to explicitly include output flow from the socioeconomic system to complete environmental impact assessment and valuation.

The delivery flow through the system process is the substance of socioeconomic life, and is a way to measure thresholds of change. Within a system, there are tolerance levels with regard to variation of deliveries. Systems respond to flows according to the level, or amount, of the flow. It is through flow levels that systems are integrated. For example, the level of aggregate demand delivered in the economy influences the level of employment. Delivery levels inconsistent with the tolerance threshold will create negative feedback for change. As examples, food deliveries may be inadequate or the air pollution level may be too great.

Negative and Positive Feedback

For policy purposes, especially with regard to the natural environment, the system concept of negative and positive feedback is very important. "Negative feedback is associated with self-regulation and goal-direction, positive feedback with growth and decay."[23] The inputs of living systems consist not only of energy and material, but also of information, all of which "furnish signals to the structure about the environment and about its own functioning in relation to the environment."[24] Feedback is a form of inter- and intra-systemic communication in which the past performance of the system yields information to guide its present and future performance. Negative feedback systems are error-activated and goal-seeking in that the goal state is compared with information inputs on the actual state and any difference (error) provides an input to direct the system toward the goal state. Negative feedback, thus, leads to the convergence of system behavior toward some goal. "When the system's negative feedback stops, its steady state disappears, and the system terminates."[25] One of the main benefits of democracy is the negative feedback and interference from the citizens, who evaluate the condition of the system and compare it to the goal desired.

What makes the open systems approach so vibrant from a policy standpoint is the fact that it views the environment as being an integral part of the functioning of a sociotechnical system. Thus, external forces that affect a system need to be accounted for in the analysis of the system. Furthermore, negative feedback mechanisms are needed to provide information about environmental changes that will affect the system in order to better understand what, if any, policies need to be made to ensure a continued effective system.

Positive feedback systems, in which positive feedback information overwhelms negative feedback information, tend to be unstable since a change in the original level of the system provides an input for further change in the same direction. "Society and technology tend to reinforce one another in a positive feedback manner, which is not always desirable. At the same time there is often a loss of negative feedback and self regulation."[26] For example, if an agricultural system based on cultivation technology is not incorporating the negative information regarding soil erosion, the system will continue its growth until destruction.

Differentiation and Elaboration

Biological and social systems' behavior is distinguished from non-living systems' behavior by the unique character of the tendency of living systems to evolve toward greater and more significant differentiation and complexity.[27]   This idea has been expressed in almost all significant disciplines. Katz and Kahn have stated with regard to social systems that "open systems move in the direction of differentiation and elaboration . . . . Social organizations move toward the multiplication and elaboration of roles with greater specialization of function."[28]   A similar evolution exists with regard to the economy. "In the economic sphere, a traditional society displays relatively little division of labor, but modern societies produce a proliferation of highly differentiated and specialized occupational statuses and roles."[29] Differentiation and elaboration become important characteristics when considering policies both in terms of anticipating changes in systems and changes that policies will engender.

Real Time

The time concept most consistent with GSA is system real time. It is inconsistent with classical ideas about time. Ludwig Von Bertalanffy explained that according to the classical Kantian system: "There are the so-called forms of intuition, space and time, and the categories of the intellect, such as substance, causality and others which are universally committal for any rational being. Accordingly science based upon these categories, is equally universal . . . . Newtonian time, and strict deterministic causality, is essentially classical mechanics, which therefore, is the absolute system of knowledge, applying to any phenomenon as well as to any mind as observer. It is a well-known fact that modern science has long recognized that this is not so."[30]

Modern science applies the time concept which is most appropriate for the subject under investigation. "The biologist finds that there is no absolute space or time but that they depend on the organization of the perceiving organism."[31]   A similar idea is found in the concept of experienced time. "Experienced time is not Newtonian.

Far from flowing uniformly . . . it depends on physiological conditions."[32]

Time is not a natural phenomenon; rather, it is a societal construct. The construct should be consistent with the GSA view and counter to the reductionist view. Time, if it is to be a useful tool for policy analysis, should be what usually is connoted by the word "timeliness." Timeliness requires that we ask the question: Which policy program will sequence and deliver the right amount of deliveries to the correct components at the right points in the socioecological system to allow for integration, maintenance, and restoration? Temporal evaluation that judges whether a project correctly sequences the delivery of impacts with system needs is consistent with the basic concepts of real time, as found in computer science. Real-time systems relate to the sequential events in a system rather than to clock time. The system itself defines when events should happen.

Evaluation

Systems are maintained through regulation, control, correction, and adaptation as a consequence of evaluation. Evaluation is common to all systems. Consistent with GSA, "analysis, evaluation and synthesis of systems is not concerned primarily with the pieces . . . but with the concept of the system as a whole; its internal relations, and its behavior in the given environment."[33] The focus of evaluation is to identify the value of the various entities as they contribute toward making the socioecological system viable. Viability includes the idea that there be redundancy in the system network and deliveries to maintain system sufficiency. Evaluation assists in making decisions about the maintenance, coordination, and restoration of systems through the coordination and sequencing of relevant events and sufficient flow deliveries.

For program and policy evaluation, both instrumental philosophy and system principles need to be applied. To complete such inquiry, measurement, data bases, and statistics are necessary. Legitimacy in numerical matters depends on social and philosophical relevance. This means that relevant facts and valid numbers are socially constructed indicators. A review of this understanding is the task of the next chapter.

CHAPTER 5

SOCIAL CRITERIA AND
SOCIOECOLOGICAL INDICATORS

The preceding two chapters were devoted to the importance of instrumental philosophy, criteria, and systems principles for policymaking. To capture the benefits of those concerns, the difficult task of generating databases consistent with their guidance is necessary. The policy analysis of the modern age both uses and directs the development of large quantities of data and information that are social indicators. The purpose of this chapter is to present a general methodology for creating social indicators. The methodology rejects the possibility of socioeconomic or ecological evaluation via any single criterion. The first section of the chapter is devoted to the conceptualization of measurement in a public policy context, with the second section devoted to particular examples.

As explained earlier, research should be context specific. This rule should especially be heeded in policy research, and the research and measurement should be consistent with the relevant context. The context is defined by the problem. "An essential question to ask of any piece of policy research is: whose 'problem' is being investigated? A 'problem' in social science can mean one of various things."[1] What we identify as policy problems is determined by our cultural values and social beliefs. Thus, the values and beliefs should be consistently applied in all aspects of the design and construction of policy research and measurement. Thorstein B. Veblen was an early proponent for binding analysis and data collection to belief criteria. In 1899, he explained in the *Theory of the Leisure Class* that: "The ground on which a discrimination between facts is habitually made changes as the interest from which the facts are habitually viewed changes .... The particular point of view, or the particular characteristic that is pitched upon as definitive in the classification of the facts of life depends upon the interest from which a discrimination of the facts is sought."[2] As was emphasized in the social indicator movement that began in the 1960s, all useful measures are ultimately social. They are recognized as social indicators to indicate that they are relevant to some social context, rather than as ultimate "measures" having universal applicability.

Kenneth Land stated that "a social indicator may be defined as a statistic of direct normative interest which facilitates concise, comprehensive, and balanced judgments."[3] Therefore, "the criterion for classifying a social statistic as a social indicator is its informative value which derives from its empirically verified nexus in a conceptualization of a social process."[4] "Social process" should be defined broadly, as indicated above in Figure 2-2. For social indicators to be completed in the area of water resources, for example, it is necessary to draw on knowledge from the disciplines of political science, geography, philosophy, ecology, economics, and engineering. Consistent with instrumentalism, it is important to recognize that policy indicators should be developed consistent with the problem, the relevant system, and the social belief criteria. This does not mean that every fact and number used in a particular program analysis has to be generated anew. Industries already created and available may suffice. However, before using an indicator that already exists in a database, it is imperative to determine whether its creation was designed and collected consistent with the needs of the current context. Likewise, analysis should not be contorted to fit a database that already exists. Does the indicator have the appropriate background, come from the correct home, have a proper upbringing, and exhibit the qualities of growing up right?

Indicator Design for Policymaking

To design relevant indicators, the measurement standards, as summarized from John Dewey,[5] should be applied. They are:

Consistent with the Problem. Indicators should be consistent with the needs of the socioecological problem being pursued. Indicators should not be recycled data collected for other purposes.

Not Necessarily in Numerical Form. Indicators are not all in numerical form. Qualification as well as quantification indicators are needed.

System Quantification. Mere separation of discrete objects is not the basis of numerical identity. Quantification should be designed to express a system.

Aggregation. Aggregation of discrete objects is not a case of measuring, but mere counting. Until a system is defined, quantification leads to indeterminate or incommensurable aggregates.

Limiting.   Social measurement must be relative and limiting—relative to the system and expressing the limits required by all systems. *Systems Characteristics.*   Systems principles of arrangement and order should guide numerical expression.   Thus, the data system should be designed to articulate patterns, sequences, ordering, and linkages. *Integrated.*   It is important to remember that, in reality, systems are not disintegrated.   Environmental conditions, institutions, and organisms exist only as a synthetic whole.   Indicators should assist in understanding and evaluating the integrated system. *Non-social Entities.*   System specification must include those indicators needed to apply physical and biological laws and their interactions, along with technology and its relationships. *Site-specific Ecology.*   System specification must also include conditions like soil, sea, mountains, and climate—the ecological system in general.   Thus, a social indicator system should be a geobased data system. Figure 5-1 (which is an elaboration of Figure 2-1) demonstrates that social indicators are designed as the secondary criteria for the more primary criteria.   The primary criteria are the social policy goals that follow from the societal beliefs, values, and ethical standards.   Fact-finding cannot be separated from beliefs and values.   "The realm of fact can be neither defined nor specified without using certain values, that it is impossible to stand firmly on the fact side of the fact-value distinction, while treating the other as vaporous, and finally, that the same processes which carve facts out of undifferentiated unconceptualized stuff also carve out the values."[6]   Figure 5-1 reflects the concept of measurement as a spectrum from qualification to quantification, as explained by John Dewey.   For example, a society with a cultural value that stresses dynamic individual action will have policy goals for good health.   Thus, to assess public health programs, it is necessary to design operational measures (secondary criteria) such as the number of hospital beds per thousand of population, the change in the disease level, and so forth.

It is important, as the economist Roland McKean clarified long ago, that the indicator be consistent with the primary goal, because *operationally* the indicator becomes the public policy decision criterion.[7] It is possible conceptually to distinguish between primary and secondary criteria, but operationally it is not.   The secondary criteria become *the* action criteria.   A primary goal of, let us say, an efficient engine, differs greatly in reality depending on whether one uses a horsepower

Figure 5-1. Policy Analysis Paradigm: Socioecological Indicators

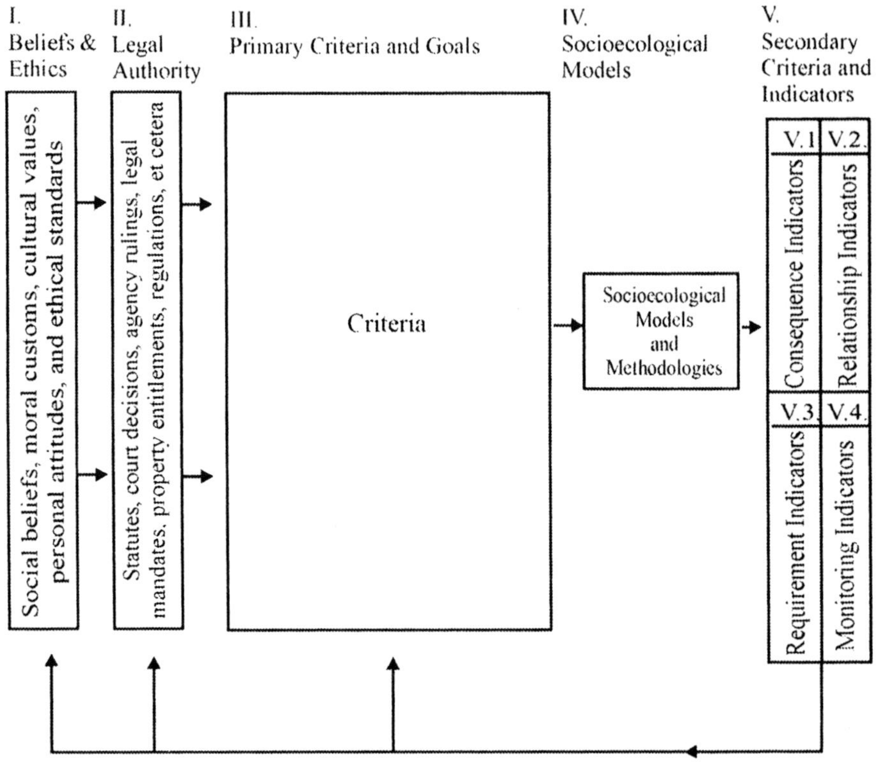

or a pollution indicator, and educational quality differs greatly depending on whether one uses an expenditure per student or a standardized test score as the indicator. In reality, the policy indicators, if applied, determine the final policy result. The U.S. Court of Appeals, District of Columbia stated in July 1989 that: "Whether a particular choice is efficient depends on *how the various alternatives are valued.*"[8] How they are valued also depends on the social indicators used for their valuation. Thus it is important to make sure the social indicators used to measure and value alternatives are consistent with primary indicators.

Figure 5-1 is constructed to demonstrate that the kind of indicators compiled depends on the socioecological model or methodology utilized. An indicator derives its legitimacy as an informative tool from being empirically verified in a model. It would therefore be necessary, as indicated in Figure 5-1, for the models and methodologies to be consistent with the primary social criteria and goals. "The social scientist's choice of problem is given exact form when he or she comes to define and specify the concepts to be used in a particular study."[9]   For example, since ecological problems have a social base, ecological models should include both social and ecological systems and so should the database of ecological issues.

Measurement and indicators, if they are to be useful, should be system indicators instead of just inputs and outputs. Thus, they need to be compared to system criteria to determine their adequacy. Indicators need to be derived in such a way as to allow for the articulation of system attributes such as structure, linkages, deliveries, and control mechanisms. If there is concern for restoration of a damaged ecological system, for example, the functioning of those system attributes is valuable for restoration and, therefore, needs to be ferreted out through appropriate system methodologies. Indicators can, from a system point of view, be categorized as follows:

Consequence indicators that are designed to measure the results of policies, or damages, or ongoing system processes.

Requirement indicators that measure the contributions to the system of the required system elements.

Relationship or linkage indicators that measure the relationships and congruency among system elements and components.

Monitoring indicators, which are selected to provide information on some part of a system, especially after policy initiatives, to determine if system value is being maintained consistent with policy intent.

In Figure 5-1, social beliefs are divided into two parts. Part I is the Beliefs and Ethics section, and Part II is the Legal Authority. Legal authority concerns have been developed consistent with social beliefs, especially as expressed by legislative bodies, and, in turn, the primary social criteria have been developed consistent with the legal authority. A listing of a set of primary indicators is contained in Part III.   The valuation indicators resulting from applied methodologies are indicated in Part IV. The categories of secondary indicators in Part V will depend on the problem and the methodologies used to generate the data.

Examples of Primary Criteria

The primary criteria listed in Figure 5-2, and elaborated below, were developed after studying sources such as statutes, court opinions, policy statements, and scientific literature. (See, for example, the Comprehensive Environmental Response, Compensation and Liability Act [CER-CLA], Superfund Amendments and Reauthorization Act [SARA], and *Ohio v. U.S. Department of the Interior 1989.*) Under the overarching goal to protect natural resources, the following primary criteria were derived from the legal documents and scientific literature for defining the restoration costs in the case of hazardous waste damage to natural resources.

Damage Assessment. To develop standardized techniques for assessing both the biological and economic damages from releases of hazardous substances.

Capture Losses. To capture fully all aspects of loss in determining damages, including both direct and indirect injury, destruction, or loss, and taking into consideration factors including, but not limited to, replacement, use value, and the ability of the ecosystem or resource to recover.

Cost-Effective. To select remedial actions that provide for cost-effective actions. The required costs include the total short- and long-term costs of such actions, including the costs of operation and maintenance for the entire period during which such remedial activities are necessary.

Non-Market Measure. To employ non-market measures for the value of natural resources because natural resources have value not measured by traditional means.

Cost is not Value. To not view market (or cost-benefit) value and restoration cost as being equal or as having equal presumptive legitimacy. Traditional means of value are not consistent with the measurement of restoration costs.

Resource Restoration. To recover all costs necessary to restore the habitat and its inhabitants to their condition before the release of the hazardous substance. For example, if the spill of a hazardous substance kills a rookery of seals and destroys a habitat for seabirds at a sea life reserve, then complete restoration is the intent; to make whole the natural resources that suffered injury from release of the hazardous substance. Such damages are to include both direct and indirect injury, destruction, or loss, and are to take into consideration factors including but not limited to replacement value, use value, and the ability of the ecosystem or resource to recover.

Replacement Cost. To recover replacement costs beyond restoration costs if applicable. The excess over restoration costs must be used to

Figure 5-2. Policy Analysis Paradigm with Primary Criteria

I. Beliefs & Ethics	II. Legal Authority	III. Primary Criteria and Goals	IV. Socioecological Models	V. Secondary Criteria and Indicators
Social beliefs, moral customs, cultural values, personal attitudes, and ethical standards	Statutes, court decisions, agency rulings, legal mandates, property entitlements, regulations, et cetera	1. Damage Assessment 2. Capture Losses 3. Cost Effective 4. Non-Market Measures 5. Cost is not value 6. Resource Restoration 7. Replacement Cost beyond Restoration 8. Use Value beyond Restoration	Socioecological Models and Methodologies	V.1 Consequence Indicators / V.2 Relationship Indicators / V.3 Requirement Indicators / V.4 Monitoring Indicators

acquire the equivalent of the damaged resource—even though the original resource will eventually be restored. This cost is to cover whatever must stand in for the injured resource while restoration is underway. Flows of services provided to the public by the resource may be curtailed long after the physical, chemical, or biological injury has abated. If a damaged forest is replanted with small trees, many years will pass before a mature forest emerges.

Use Value. To recover interim use values beyond restoration if applicable. The measures of damages must not only be sufficient to cover the intended restoration or replacement uses in the usual case, but may in some cases exceed that level by incorporating interim lost use values of the damaged resources from the time of the release up to the time of restoration. Use value is to be limited to "committed use," which means a current public use or a planned public use. This avoids the need for unreliable, and likely self-serving, speculation regarding future possible uses. Option and existence values are included as use values.

To accomplish the goals elucidated by these primary criteria, numerous measures need to be developed. Several aspects of wildlife habitat defy market valuation, and information regarding the value of habitats is necessary to take full account of the impact of regulations and policies on the environment. Neither one measure nor one category of measures is sufficient to express or value system goals, nor can any one measure or concept serve as a common denominator for all the diverse indicators required.

This point can also be seen with the example of primary criteria for state and local pension funds in the United States. The primary criteria as listed below are legally established to accomplish an array of social beliefs as established by the relevant political communities. The relevant criteria are:

Universal Coverage. To provide for universal coverage because as a member of the community an individual has a right to an old-age pension.

Minimum Pensions. To provide a minimum pension irrespective of work or earning history. Those individuals with inadequate or no work and earning history have a right to a minimum pension.

Additional Source of Income. To provide additional income to particular employee groups in the form of pension plans instead of wages.

Continuity of Pension Income. To mitigate the uncertainties of pre-retirement and retirement life by the maintenance of a continuous flow of income during retirement.

Adequate Retirement Income. To maintain an adequate retirement income.

Purchasing Power. To adjust the pension to a price index in order to maintain real purchasing power after inflation.

Maintenance of Relative Socioeconomic Position. To adjust the pension for inflation, productivity, and other changes in order to maintain the relative socioeconomic position of the retired individual.

Intergenerational Equity. To maintain a pension that bears a certain relationship to pre-retirement earnings for each member in the same generation.

Redistribution Impacts. To redistribute income and wealth across income groups.

Intergenerational Transfers. To redistribute income and wealth across generational groups.

Influence of Corporate Governance. To sponsor shareholder campaigns to influence corporate governance issues (such as Delaware incorporation, poison pills, confidential voting).

Macroeconomic Impacts. To influence macroeconomic variables such as savings rates, employment, economic growth, investment, interest rates, or labor markets. This might be referred to as economic development or regional economic impacts. Economic development is one of the many reasons pension fund managers pursue program-related investments. For example, many states have geographically targeted venture capital funds for small business startup and expansion.

Beneficiaries' Vital Interests. To direct pension investments so they do not impair the vital economic and social interests of the fund beneficiaries. Beneficiaries' vital interests are a concept similar to that of macroeconomic impacts. The difference is that the impacted party is more narrowly defined and the consequences of concern are more broadly defined. The party is limited to beneficiaries—not the broader community—and the list of concerns goes beyond economic impacts to include health, wage scale, labor relations, pollution, and so forth.

Governmental Financial Stability. To invest in the bonds of governmental units that sponsor the pension. Since the pension plan is dependent on the fiscal health of its sponsoring government, the pension funds are sometimes used to buy government bonds in a time of fiscal crisis.

Social Investments. To direct investment expenditures in a manner so as to increase production of particular goods and services. Housing has been one of the main recipients of social investment.

Social Responsibility. To direct investment into organizations whose organization, finance, production, and marketing are conducted in a manner consistent with accepted community and ethical standards, or to encourage the management of organizations in which funds are invested in a socially responsible manner. There are numerous social responsibility criteria.

Positive Geographical Targeting. To direct investment into a particular geographical area, such as the fund's own political jurisdiction, in order to improve the area.

Negative Geographical Targeting. To prevent investment from flowing to certain geographic areas or political jurisdictions in order to change social, financial, economic, racial, or corporate policies in the area.

High Technology Development. To achieve advanced technology developments and innovation through investments in research, seed capital, and high technology investments.

Appropriate Technology Development. To direct funds into plant and equipment investments that meet appropriate technology criteria.

Environmental Protection. To target investments to protect, conserve, restore, or regenerate the natural environment.

Every criterion on the list is not applicable to all state and local pension funds. In fact, some are inconsistent with others on the list. However, all are applied to some of the pension plans. The point of interest here is to demonstrate that different primary criteria will call for different models for analysis and, therefore, to be useful when making decisions require different indicators to understand and evaluate whether societal goals are being reached through investment and distribution activities. Some primary criteria call for models of financial markets and financial risk indicators such as beta coefficients in order to determine whether there is an adequate return on stock investments and to balance the expected return with a prudent level of risk. Other primary criteria for pension investment call for models and indicators related to ecological systems.

Interesting research documents for observing the issue at hand include (1) *Social Security: Criteria for Evaluating Social Security Reform Proposals* and (2) *Social Security: Evaluating Reform Proposals*, which were completed by the United States General Accounting Office (GAO). GAO is responsible for completing research on public policy issues as assigned to GAO from Congress. Given the controversy surrounding the Social Security issue, when the request came to GAO to analyze the potential budgetary and economic effects of Social Security reform, there were several different contending plans. They were (1) the Social Security Guarantee Act outlined by Ways and Means Committee Chairman Bill Archer and Social Security Subcommittee Chairman Clay Shaw, (2) HR 1793, the 21st Century Retirement Security Act, (3) the Senate Bipartisan Social Security bill proposed by Senators Judd Greg, Robert Kerrey, John Breaux, and Charles Grassley, (4) the Social Security plan outlined by Budget Committee Chairman John Kasich, and (5) the Social Security framework outlined by President Bill Clinton. These proposals came from different ideological belief perspectives. Since research has to be guided by social beliefs, it must be determined what set should be utilized given the very different criteria being expressed by the different proposals. GAO, first, arrived at three general criteria to provide guidance for general research work.   Next, sets of more refined and detailed criteria were determined for each of the general criteria; and, finally, additional criteria were added that were unique to each particular Social Security plan as each plan was analyzed.   Thus, the analytical framework for policy assessment was guided by carefully designed social criteria relevant both to the general interest and to particular interests.

Conclusion

Over the years, various groups have proposed various indicators to serve as the single measure or the common denominator function. These have included monetary prices, BTUs, protein ratios of the food chain, hours of leisure time, and so forth.   Each of these failed to meet such an impossible standard.   The failure of BTUs even as a measure of an energy system can serve as an example.

Not all forms of energy are the same.   Some forms of energy such as nuclear fission, electricity, or gasoline are quite concentrated or of high quality.   These forms can perform a lot of useful work per pound or cubic foot of material.   Other forms, such as sunshine, tides, wind, and low temperature heat, are somewhat diluted and spread out over a large surface or volume. These forms do not have much useful work to offer, even though the total amount of energy might be the same as for a more concentrated form.   Thus, in combining and evaluating the contributions of various systems, it is important that equivalent forms of energy be used.   This is analogous to the old saying that we cannot add apples and pears.   Likewise, we cannot add sunshine BTUs or kilocalories to gasoline BTUs or kilocalories and expect the total to accurately reflect the amount of work that can be done by that energy.[11]

Like BTUs, all dollars are not of the same value, so they cannot necessarily be added to determine cost.   Equal dollar cost figures that show equal expenditures for flu vaccinations and cigarettes do not indicate the same value or the same burden.   Thus, it is important for policy scientists to develop methodologies that will allow for the creation of the needed indicators consistent with social beliefs and primary goals.

CHAPTER 6

THE SOCIAL FABRIC MATRIX

The social fabric matrix (SFM) is a methodology based on theoretical and technical developments from numerous areas. It has been developed to allow for the convergence and integration of conceptual frameworks in instrumental philosophy, general systems analysis, Boolean algebra, social system analysis, ecology, policy analysis, and geobased data systems. The focus of the SFM is to provide a means to assist in the integration of diverse fields of scientific knowledge, utilize diverse kinds of information in order to describe a system, identify knowledge gaps in a system for future research, analyze crises and opportunities within the system, evaluate policies and programs, and create social indicators for future monitoring.

The SFM allows for the application of two principles explained by John Warfield. He wrote, "first, there is the *principle of association*, which states that the developer of a model must engage in associating elements of representative systems with those things that are to be modeled. Second, there is the *principle of model exchange*, which states that it is desirable to find ways of transforming a model from one representation system to another to meet the needs of understanding, learning and effective communication."[1]   The SFM allows for those principles to be applied for complex systems. For real-world modeling, Clifford Geertz stated that "explanation often consists of substituting complex pictures for simple ones."[2]   Thus, as stated above, he advised to seek complexity and order it.   Modeling complex systems "deals with how the human, that is, behavioral and social subsystems affect and are affected by the nonhuman, that is, the technological subsystem, and how these subsystems collectively in turn affect and are affected by the usually dynamic social and natural environments in which the larger system is enmeshed."[3]

The SFM has been developed because, as the scientific literature and policy experience indicate, a narrow conceptualization of socioecological systems is not viable. As ecological concerns have grown, it is clear that socioeconomic systems are open systems. This is especially recognized with regard to economic processes.

> In fact, economic processes can be understood only as depending upon a continuous "exchange" of energy and matter between the economy and nature. In the course of these largely non-market exchanges, available or economically accessible energy/matter are transformed first into inputs and then into vendible outputs and partly into residuals which will be dispersed into the atmosphere, the water, and the soil, giving rise to qualitative and quantitative changes of both the environment and the economy itself. . . . Hence economic processes have the effect of continuously altering the conditions of the environment and the economy. . . . These changes of the environment and the economy may become cumulative with far-reaching negative consequences for the conditions of human health and life, and may even endanger the condition of economic and social reproduction in the long run.[4]

Therefore, the economy can neither be understood nor analyzed by a simple modeling apparatus, it requires methodologies that allow for the interaction between the ecological system and social institutions.

The SFM serves as a means to organize and structure policy research. In addition to Albert Einstein's tremendous substantive contribution to physics, he pointed out that the results found in scientific investigations depend, even in physics, on the frame of reference and view of the investigators. This knowledge has had a pronounced impact on scientific methodology. This is true in all sciences, but it is especially true in the policy and decision sciences. Therefore, in order for research to be relevant to the problem, it is necessary to structure policy research consistent with the decision maker's frame of reference and primary criteria. Policy research and information models can be designed in order to encourage researchers to ask the right questions and compile information in order to answer them consistent with the relevant frame or context. In this way, diverse technical expertise can be harnessed into a unified system to strengthen evaluation and decision making. The SFM can be utilized as a tool for organizing policy analysis for complex systems because it allows for the comparison of alternative systems that result from alternative policy scenarios.

A task of paramount importance in policy analysis and program valuation is the comparison of alternative systems. To know and understand a problem, it is necessary to model and articulate the system that is creating the problem. To make a decision about whether a policy and/or program(s) will improve the situation, it is necessary to model and articulate the new system process that will function after the new policy is implemented. This will allow for the two systems to be compared to decide, prior to implementation, whether the new system created by the new policy is superior.   More than one alternative should be researched, if resources allow, so that more than one policy alternative can be compared.   The SFM is a means to compare and evaluate alternative systems created by alternative policies and their programs.   Too often, new programs have been implemented without system analysis only to discover, by the experience of undesirable consequences, that the new programs made the system worse.

Components of the Social Fabric Matrix

The literature from anthropology, social psychology, economics, and ecology suggest six main components that need to be identified and integrated in order to understand a problem and to make policy to solve it.   As stated above, those components are (1) cultural values, (2) societal beliefs, (3) personal attitudes, (4) social institutions, (5) technology, and (6) the natural environment. Before it is possible to assemble a framework for defining the relationships among these components for a problem, it is necessary to understand the components. We begin with cultural values.

Cultural Values

There are numerous meanings the word "values" evokes; for example, social value, instrumental valuing, technological values, valuation, and so forth. Those concepts are usually concerned with "what ought to be," while the definition of values set out here is intended to describe "what is."

Values are a subset of culture. A culture is a collective systemic mental construct of the superorganic and supernatural world. It does not exist as a whole in any single mind. It contains a group's abstract ideas, ideals, and values from the superorganic and supernatural world, and it is found in legends, mythology, supernatural visions, folklore, literature, elaborated superstitions, and sagas. Culture is provided by tradition and not by the human agent or social institutions. This means culture is not determined by instrumental valuing, discretionary decisions, or technological change. Culture is separate in definition, meaning, and performance from society. Culture, although a powerful directive and prescriptive influence on society, is cerebral, while society is the set of sociotechnical relationships that direct behavior patterns. Society changes regularly but culture does not; a culture lives on even after a society is destroyed. The culture is the center. Samuel P. Huntington recently applied the concept of culture, using the distinction between culture and society, to explain the long-term clashes and on- going terrorism to be expected among the Western, Islamic, and Sinic cultures. His important conclusions indicate the importance of the concept of culture for real-world policy.[5]

Cultural Values as Criteria

Values are cultural criteria or evaluative standards for judgment with regard to what is ideal. They are the ultimate criteria in the sense that they are above social institutions. They are the focal criteria that are the locus to which all social criteria attempt to conform. To achieve a well-adjusted society, instrumental policymakers should design policy criteria and institutional patterns to be in conformance with cultural values.

A problem regarding the analysis of cultural values is that scientists often attempt to apply the concepts, which have been developed for analyzing society, to analyzing values. For purposes of analysis and relevant policy, we need to distinguish cultural values from other entities that are sometimes referred to as values. Cultural values are not desires, motives, pleasures, beliefs, attitudes, or tastes. Neither are they determined by instrumental valuing and discretionary decisionmaking, nor by technological change. Neither do they conform to theories of value in economics nor to monetary exchange. Neither are they prioritized along some hierarchical cardinal or ordinal scale nor measured by price. Neither can they be added up as production nor functionalized as a "social welfare function." The concepts just mentioned are concepts taken from societal analysis. They are not appropriate for the analysis of cultural values.

The analysis of a group's stories, holy books, legends, and so forth reveals basic criteria common to all those sources. Those same criteria guide the group's social relationships. While the basic cultural values are limited in number to a dozen or so, the belief criteria guided by values number in the thousands.

Western values that have been the same for centuries include (1) strong domination of nature by humans, (2) atomistic conceptualization, (3) extensive hierarchical relationships, (4) flowing time, (5) dualistic thought, (6) dynamic expansiveness, and (7) universalism. They are found in our legends, songs, religious myths, and literature, and are acted out in our social arrangements. Although these values have been constant for centuries, they have been expressed through different societal arrangements in different eras. Cultural values call forth different social beliefs, rules, and regulations in different historical situations.

When folk tales and literature are analyzed, it is found that Western heroes and gods operate in a system consistent with the above-listed cultural values. Authority and power are distributed over a wide hierarchical range (kings, queens, and barons; corporate CEOs, vice-presidents, and comptrollers; chancellors, professors, and assistant professors; and cardinals, bishops, and priests). The agents operate in a dynamic world where the existence of growth and creation is good (amount of empire taken, growth of sales revenue, number of articles published, and souls saved); where the natural environment exists to be subdued; (slaying dragons, damming raging rivers, and draining evil swamps); where dualistic structures are clear (good and evil, heaven and hell, labor and capital, demand and supply, and profit and loss); and where the faster it is all processed, the better (it is a sin to waste passing time).

Cultural values are transcendental across all aspects of culture and society. They assert themselves even in the unconscious spheres of existence. Because they are transcendental, it is impossible for the human agent to change them. They are constantly being reinforced through culture, in the workplace, by movies, television, language, linguistic structure, and so forth. Their reinforcement is direct in the culture and indirect through beliefs in society. The values are the evaluative criteria for establishing which actions and relationships are worthy of providing satisfaction and which should be desirable. Cultural values are not goals or actions, they are end-existence criteria by which goals and actions are to be judged. They are the basic and primary prescriptive forces that circumscribe societal norms, which in turn serve as criteria for institutional patterns. Although powerful and transcendental, cultural values are not deterministic because numerous alternative beliefs and institutional arrangements can satisfy a set of criteria. They limit and exclude but do not determine in a specific sense.

Because there has not been adequate distinction made between culture and society, not enough work has gone into designing indices to determine whether basic cultural values are being met. Given their importance, they should be ignored least of all. This is true for at least two reasons. First, resources can be wasted if the economy is directed by trifles. Distractions such as pleasure, interest, or compulsions may misdirect energies and resources outside the spectrum of activities that culminate in worthy activity patterns. If that happens, dissatisfaction will set in and the abandonment of established economic structures will be expensive. Second, a lack of concern for cultural values leads to alienation with its accompanying psychological problems, social malaise, and loss of economic productivity.

Just as we accept the scientific fact that goose down is softer than door knobs, and therefore do not use the latter to stuff pillows, we need to accept anthropological findings that there are cultural criteria that must be taken as given. Thus, if Western culture has a strong emphasis on dominating nature, Westerners cannot solve environmental problems by designing programs to live in harmony with nature. Instead they should design programs that allow for domination without adverse repercussions. For example, there are ways to cultivate the soil that cause high rates of soil erosion, and there are ways to cultivate that result in building the soil. Both allow for humans to express the domination trait; however, the latter does it in a manner that serves civilization.

Although anything short of genocide will not destroy a culture, an inability to express it will lead to a deteriorating condition of any group. The historical case of native peoples under colonial rule is an example of extreme deterioration. Because these people were not able to express their cultural values in their social and economic activities, their condition became a deteriorating one of alienation, alcoholism, and crime. If the modern society is not to create for itself what was created for native peoples, efforts should be devoted to the development of value indicators to be used for the enhancement of cultural welfare.

Social Beliefs

Whereas cultural values are transcendental, social beliefs are activity and institution specific. The connection between values and beliefs (as indicated above in Figures 2-2 and 3-2) provides the bridge between culture and society. Society is a set of relationships; not people, or bronze, or horses. The relationships are determined by institutions, which are patterns of activity that prescribe the roles for the elements (humans, animals, machines, trees) as well as the emotional commitments for the human element.

Some analysts have attempted to convince us that social institutions are consequences of rational valuing. Not only is that lacking evidence with regard to the vast majority of institutions, in addition, we do not want to fall into the trap of assuming there is some deterministic process that automatically selects optimal social institutions. If that were the case, policy analysis would be unnecessary.

Walter Neale stated in his explanation of social institutions that an institution is identified by three characteristics. The characteristics are: (1) patterns of activity, (2) rules giving the activities repetition, stability, and order, and (3) social beliefs explaining or justifying the activities and the rules.[6] The answers to the questions about "why" institutions function as they do "reflect the beliefs of the participants about how and why the activities are carried on *or* beliefs about what justifies or ought to justify the activities."[7] To summarize, institutions are the repetitive patterns of activity which contain the roles of the elements, provide for the structure of societal relations, and direct the flows of societal substance. Institutions are found in society, not in culture. The prescribed and proscribed institutional roles of the elements of society are based on rules of prohibition, obligation, and permission. These rules are a consequence of the norms and beliefs which evolve from and are enforced by the social process. Normative beliefs permit or prohibit the doing of an action and the necessary connections that make (or do not make) the doing or forbearing of the action a practical necessity.[8]

Since institutions are accepted as normal behavior, the belief criteria justifying them are referred to as norms. These normalized beliefs are the societal criteria for what is good and bad, correct or incorrect, and are, in a stable non-alienated society, in conformity with cultural values. Each institution and activity will have a cluster of beliefs that are specific to that institution. Each belief conforms to all cultural values. An ideology is a system of congruent societal beliefs. Thus, in analyzing a social or economic problem, ideological analysis is very important.

To determine efficiency, it is necessary to consider whether institutions and economic processes fulfill cultural values and societal beliefs. Maximizing production levels or optimizing around production functions and consumer preferences are not the core efficiency concerns. Efficiency is a matter of meeting social criteria. If this is understood, the decisions on the part of courts and governmental agencies regarding job safety, hiring quotas, and hostile corporate acquisitions would not be considered external interferences. They are efficiency concerns from a social point of view.

Searching out belief criteria is imperative for policy analysis. In a modern society, beliefs are usually expressed through codification in statutes, agency rules, court decisions, regulations, and legal opinions; and legal criteria are established for judging everything from university hiring procedures to water quality. Social beliefs are the battle zone between culture and society when there is social change. Since beliefs give rise to institutional means of expression, cultural values are, therefore, made obligatory on societies' members by institutional arrangements. If social institutions are changed for some reason, let us say because of technological innovation, new beliefs will develop consistent with the new pattern of behavior. The new patterns and beliefs may be inconsistent with the unchanging values. The inconsistency leads to alienation and social disorientation, which in turn leads to new social processes. This is usually a haphazard trial-and-error process. This is one of the reasons the social fabric matrix framework is needed to assist in social alignment for the development and legitimization of new institutions.

Social beliefs and institutions establish roles for the elements. For each institutional situation there are obligations, permissions, and prohibitions for the elements. The human element is socialized to respond to signs and symbols in order to fulfill the responsibilities and duties of a situation. These responses are referred to as attitudes. Beliefs and institutions regulate peoples' attitudes (responses) toward signs and symbols, and thereby regulate behavior. "The argument is that person, situation, and behavior all affect each other continuously."[9]

Attitudes as Human Responses

Attitudes represent several social beliefs focused on a specific object or situation. It is through attitude responses that the machines are minded, the children get fed, the flags are saluted, and the trees are cut. It is through attitude theory that the human actor is brought into socioeconomic theory. Attitudes are held by specific people, about specific objects and situations.

After hedonism and instinct theory fell into scientific disrepute, inner drives and motives were postulated as the mechanisms from within the human that arouse, direct, and sustain activity. The *Dictionary of Behavioral Science* has defined a motive as "a state within an organism which energizes and directs him toward a particular goal."[10] Reductionists assumed they could reduce to motives and then build up a social system by aggregating motives. This approach has been rejected in the psychological sciences. Time and time again attempts to study attitudes through introspection were lacking in verification.[11] The reductionist approach has been denied. The claims of utility calculation and hedonism "when tested in the crucible of social policy, proved inadequate."[12] The scientific reliability of motives was soon questioned even for studying hunger, thirst, and the sex drive.

With the development of social psychology, the idea that motives were reliable operatives continued to lose credibility. The *Encyclopedic Dictionary of Psychology* states that "in the early days of behavioral science, motivation was envisaged in terms of the drive that was necessary for the manifestation of behavior: sexual behavior was due to the sex drive, eating to the hunger drive, and so forth. This is no longer a prevalent view and it is generally recognized that it is not necessary to account for behavior in terms of motive force."[13] Today the concern is with the role of social institutions in determining attitudes. "Attitudes are individual mental processes which determine both the actual and potential responses of each person in the social world."[14]

Psychological scientists are more interested in the mental processes while policy scientists are more interested in the responses and their origin in the social system. Buyers respond to price changes. Workers respond to coefficients coming across a computer screen on an assembly line in a factory. Bank tellers respond to customers approaching their window. Parents respond to a child's misbehavior. Workers respond to safety devices in the meat packing plants. It is such responses that articulate the socioeconomic system. The attitudes originate from outside the individual, not from motives, or hedonistic urges, or utility. Social psychologist, William J. McGuire, in the *Handbook of Social Psychology*, stated "institutional structures have intended or unintended impacts on attitudes by determining the stimulus situations to which the person is exposed, the response options available, the level and type of motivation, and the scheduling of reinforcements."[15]

No object has meaning without reference to clustered social beliefs and attitudes. However, beliefs are determinants rather than components of attitudes. A belief concerning the equal treatment of persons influences human attitudes and responses toward particular persons such as minorities, males, and the handicapped. "Put another way, the objects and situations we encounter have meaning for us not only because of the attitudes they activate within us but also because they are perceived to be instrumental to realization (or to stand in the way of realization) of one or more social beliefs."[16] It is through the day-to day attitudinal responses to signs and symbols that the person-to-person, person-to-technology, and person-to-environment relationships are maintained in an institutional arrangement.

The more affluent a society in money, technology, and information, the greater the risk that members will lose sight of basic beliefs in forming their attitudes. All three lead to change and have the tendency to make society more complex and to add additional layers of authority. Everyday activities mold attitudes, and those everyday activities may have become inconsistent with basic beliefs and values. Without effective monitoring with social indicators, human actors may misinterpret the correct response that the institutional norms are conveying. "In a traditional society, where the institutions are often stable over long periods, there is usually a close fit between cultural values and institutional actions. But in modern society, where institutions change rapidly, it can sometimes happen that the values and goals of a society are in conflict with institutional requirements."[17]

Since attitudes are crucial in determining action, it is fortunate for policymaking and social planning that they can be changed easily. "Attitudes, while important and generally resistant to change, nevertheless are of less connective importance to society and easier to change than the central beliefs."[18] While basic social beliefs are very difficult to change, cultural values are unchangeable for policy purposes. The more transcendental the concept, the more social entities there are for expressing, reinforcing, and maintaining it and, thus, the greater the connective importance.

Tastes as Unimportant Attitudes

Commodity tastes are treated here as a special category of attitudes because they are related to the institution of market demand. Although tastes have been given a lofty status in the neoclassical tradition of economists, they are the least important of the attitude categories because they can be changed easily—usually with no impact on basic beliefs—and, therefore, are without serious consequences for the belief system or the social structure. This does not mean that tastes cannot have a profound effect. For example, tastes can have a deleterious effect on human health. The point is that those tastes can be changed without a disruptive effect on the belief system.

Cultural values deliver criteria to beliefs and receive information from beliefs with regard to whether there is alignment between the two. Therefore, values deliver only to beliefs. Beliefs deliver criteria to institutions. Institutions deliver social information to beliefs to determine whether the institutions are in conformance with the beliefs. This informational connection is measured by social indicators. Banking institutions can be used as an example. An array of customary, legal, and judicial belief criteria is established for the banking industry. The industry uses those in designing its structure, process, and procedures and provides directives to form the attitudes of its employees. Attitudes are formed by directives from banking institutions and provide responses to the elements within banking organizations.

Technology

Technology, as defined above, is the combination of tools, skills, and knowledge. This is a much narrower meaning than when technology is defined as "organization" or as "technostructure." The purpose of broader definitions is to emphasize the role of technology in the establishment of societal, economic, and environmental patterns. Once selected and adopted, technology becomes woven into the social fabric "in a fashion as to build its own necessity."[19] The more narrow definition is used here because the combination of tools, skills, and knowledge is treated as only one component among all the components (see Figure 2-2). Broader definitions, such as technostructure, lump all components together without providing the detail of how they are integrated. The importance of technology in a system structure is that it has a pronounced effect on production requirements, social relationships, and the environment. A change in technology requires a change in institutional relationships, and, thus, a change in social beliefs. Those changes in turn change the inputs from and outputs to the natural environment. To understand a system and make technological policy for it, the specifics of the relationship established by alternative technology need to be modeled.

Ecological Systems

The ecological system is an open transactional construct as has been understood since the famous philosopher Alfred Whitehead used the ecological system as a vehicle for explaining holistic philosophy. The relationship of the ecological system to the social and economic institutions was outlined above in Figure 2-2 and discussed in Chapter 5. It is amazing that the connection between economic institutions and ecological systems could have been ignored so long by economists and ecologists. Nothing has ever been produced without the use of substance from the ecological system and waste products being created for the ecological system. Yet, economists generally continue to labor under the illusion of the Cobb-Douglas production function with only the inputs of capital and labor and a single output without pollution. Likewise, no ecological system exists today that is not impacted by and connected to socioeconomic institutions. Isolated ecological systems no longer exist. Yet, ecologists continue to complete models of wet-lands in rural agricultural settings, for example, without any indication of the impacts from agricultural institutions; impacts such as eroded soil, pesticides, and herbicides that regularly degrade the wetlands. In terms of policy models, the services from living systems of the ecologi-cal environment are of paramount concern. Services available from living systems include purification of water and air; regulation of at-mospheric chemistry; storage, detoxification, and recycling of human waste; natural pest and disease control; regulation of the chemical compositions of oceans; maintenance of soil fertility; and nutrient stor-age and recycling. The ecological environment is probably the most difficult category to conceptualize and define as a separate component because humans, their society, and society's economy and technology are so dependent on living systems, with both flora and fauna embed-ded—sometimes to extinction—in the social and technological proc-esses.

Integration of Components

The components, although separated above for definitional purposes, are in fact instituted in many interdependent, transdependent, and recur-rent ways. To integrate the six components in the SFM, two concepts are emphasized.

One concept deals with flow levels. Classical models empha-size rates and derivatives, but models of whole systems emphasize the integration of flow levels. Measures of flow levels are needed to fully describe societal and environmental processes. The flows of goods, services, information, people, and ecological substances through the network structure and maintain regional and community relationships. For example, a large regional bank, as it directs credit flows, helps structure the various communities in the region. The flow of invest-ment to particular kinds of cultivation technology helps determine the level of organic matter in the soil and, thus, the level of soil erosion and downstream pollution.

Another concept being emphasized for SFM integration is that real-world systems depend on delivery among the component parts. Systems deliver bads and disservices as well as goods and services. Natural environments deliver floods as well as nitrogen-fixing bacteria.

Factories deliver pollution as well as output for sale. The continuity of a system depends on delivery among components according to social rules and ecological principles. For example, income must be delivered to households for the maintenance of the economic system, and organic residue and amino acids must be delivered to ammonia-producing bacteria for the maintenance of the nitrogen cycle. Problems are created in systems when the delivery among the components is inconsistent with the maintenance of the system. Too little income delivered creates a recession and inappropriate farming practices prevent soil bacteria from surviving.

The SFM is based upon the concept of social components receiving flows from and delivering flows to other components, thus, emphasizing process. "Process suggests analysis in terms of motion."[20] A delivery is used to create another delivery. Any event will perforce need to be traced through the system to find additional linkages and flows.

Kind of Matrix

The SFM is an integrated process matrix designed to express the attributes of the parts as well as the integrated process of the whole. The matrix process is expressed in Figure 6-1. The rows identified by i represent the components which are delivering, and the columns identified

Figure 6-1. Noncommon-Denominator Process Matrix

	$j_.$	$j_.$	$\cdots$	$j_.$
$i_.$				
$i_.$				
$\vdots$				
$i_.$				

by *j* represent the components which are receiving. The SFM is a non-equilibrium, noncommon-denominator process matrix. In this matrix, which is read from left to right, the rows and columns are the same entries and are in the same order. The cell *i=j* defines what the *i*th entry (system component or element) is delivering to the *j*th entry, thus, what the *j*th is receiving. The terms "delivering" and "receiving" convey the basic idea that the process is ongoing.

The initial objectives in employing the matrix are to organize the scientific knowledge base, to use the matrix as a thinking tool, and to discover components and delivery linkages not yet recognized. Thus, research begins by accumulating a broad knowledge base, to include field observation, of the problem being studied. The first step when researching the problem is to construct a list of the main components, and elements of the components, that make up the real world. The same list of components is listed for the matrix rows as arrayed across the columns. (See Figures 6-2 and 6-5, as examples.) What is immediately found with respect to any problem is that many of the separately listed components affect each other. A row component can be followed horizontally across the matrix to discover the direct columns to which it makes deliveries based on research evidence available. Some of the deliveries will be qualitative and some quantitative; they include criteria, court rulings, pollution emissions, goods, services, and so forth.

The SFM becomes a tool to aid thinking and organizing research. As research is conducted cell-by-cell across each row, linkages among components and elements will be discovered that otherwise would have been overlooked. This process helps to discover research gaps, identified as particular matrix cells for which there is inadequate information. Furthermore, the process of completing the research for the matrix will jog researchers' memories of additional components to be added to the original list. They can quickly be inserted as new row and column entries and their deliveries noted in the cells.

As a review and preview, some specific characteristics of the SFM are:

(1) The matrix is based on the concept of delivery and process. A process is maintained   through delivery. Electric companies deliver energy, farmers deliver corn, and some industries deliver carcinogenic waste.

(2) The components listed on the left are delivering to those listed

across the top.

(3) It is a noncommon-denominator matrix without common flow properties; for example, it can handle energy, pollution, and dollars as well as water, steel, and belief criteria. It is necessary to develop many different kinds of numerical modalities in order to capture the essence of the various flows and relationships. This means that standard matrix algebra is not appropriate for the matrix, and that all the information in the rows and columns are not summative (as in an input-output matrix).

(4) The empirical observations contained in the cells of the matrix are the flows of the system. Only the direct deliveries to each cell are to be recorded. This is necessary in order to prevent the double counting that would occur if both direct and indirect deliveries were recorded and counted.

(5) The number and kinds of component entries in the rows and columns of the matrix will depend on the problem being studied and the policy analysts' interests. For example, if the problem deals with antitrust and competitiveness of the economic structure of the fertilizer industry, a few broad natural environment categories may be sufficient. However, if the problem is the impact of commercial fertilizer on nitrogen cycles, numerous refined ecological system entries will be needed to understand the relationships of the nitrates to micro-organisms, and so forth.

(6) The SFM approach defines the system as it exists; thus, the concepts of equilibrium, harmony, or wants being satisfied are not forced into the system if not relevant. Toxic waste lagoons can be delivering pollution to the water aquifer, police can be delivering arrests to individuals, and industrial processes can be delivering cancer to workers. Such deliveries are neither harmonious nor want satisfying, although they are part of the system.

(7) Most importantly, the matrix allows for model building and data collection consistent with theory. This means many of the cells in the matrix will be empty. As outlined in Figure 2-2 and explained above, attitudes and technology do not deliver directly to the ecological system and beliefs do not deliver directly to technology. Consequently, all the cells in the SFM between such components will not have entries.

Cellular Information

The hypothesized SFM in Figure 6-2 can be utilized for examples of cellular information (although the elements in Figure 6-2 are generally defined too broadly to be useful in articulating a real-world system).

(handwritten note: — use fig 2.2 to help find where "1" can't go)

Figure 6-2. Hypothetical Social Fabric Matrix

Delivering Components	#	Cultural Values				Social Beliefs				Personal Attitudes				Ecological System				Tech- nology		Social Institutions				
		Dominance Over Nature	Dynamic Expansiveness	Dimodal Duality	Egalitarianism	Work Ethic	Property Rights	Affirmative Action	Allure of Bigness	Soil Conservation	Gun Procurement	Racism	Commodity Tastes	Forest	Land	Animals	Water	Tool & Skill I	Tool & Skill II	Kinship Groups	Courts	Government	Industry I	Industry II
		1	2	3	4	5	6	7	8	9	10	11	12	13	14	15	16	17	18	19	20	21	22	23
Dominance Over Nature	1					1	1	1	1															
Dynamic Expansiveness	2																							
Dimodal Duality	3																							
Egalitarianism	4																							
Work Ethic	5																							
Property Rights	6																							
Affirmative Action	7																							
Allure of Bigness	8																							
Soil Conservation	9																							
Gun Procurement	10																							
Racism	11																							
Commodity Tastes	12																							
Forest	13																						1	
Land	14																							
Animals	15																							
Water	16																							
Tool & Skill I	17																						1	
Tool & Skill II	18																							
Kinship Groups	19																							
Courts	20																							
Government	21																				1		1	1
Industry I	22																1				1		1	1
Industry II	23												1										1	1

(handwritten note at bottom: ✳ Dark cells = can't equal 1)

The cells in Figure 6-2 that are shaded cannot have deliveries, as explained earlier in this chapter and with Figure 2-2. The cells are given an (i,j) designation which means the ith row and the jth column. Explanatory comments on particular cells are as follows:

Cells (22,22), (22,23), and (23,23) These cells are laid out as the standard industrial input-output (I\O) matrix. Although the layout is the same, several differences exist. First, it is apparent that inter-industry transactions designated by these cells are a minor part of the total process. As will be demonstrated next, entities outside the I/O table cells must be delivering for the I/O industries to function. Training and education, for example, is provided from the government to families for the delivery of skills before factories can operate.

Cells (21,19), (21,22), and (21,23) To structure industry, the government must provide legislation.

Cells (13,22) and (23,13) Lumber is delivered from the forest as industry delivers the harvesting technology to the forest.

Cells (17,22) Technology delivers criteria and requirements for structuring the production process.

Cell (22,16) Industry delivers pollution to the water.

Cells (1,5), (1,6), (1,7) and (1,8) Cultural value criteria are delivered to beliefs.

Social institutions, technology, and environmental elements do not exist as a unified whole without the guidance and emotional commitment provided by values, beliefs, and attitudes; and in turn, values, beliefs, and attitudes cannot be expressed and, therefore, kept alive without a viable system that expresses them. As is evident, no cell is an island. Numerous cells in a sequence are processing in order to deliver tractors to the field, healthcare to the public, or nutrients to wildlife. That sequence has stability and dependability because of the instituted process which can be expressed in the SFM. Understanding the organization of a system requires understanding how much, how, when, and where particular ordering relationships are imposed.

System Sequence: Boolean Matrix and Digraph

After the identification of deliveries in the cells, the matrix can be used to define the system sequence through Boolean algebra manipulation. To convert the matrix to a sequence digraph, each cell in the matrix in which there is a delivery is labeled as 1 and each cell with no transaction is labeled as 0. The matrix can be converted to a Boolean digraph (directed graph) such as represented by the simple digraphs in Figures 6-3 and 6-4. Each node (circle) in the digraph represents a row and column entry in the matrix and each edge (line) represents a cell delivery. The digraphs illustrate the sequential structure of the system and provide a picture that is easier for many to comprehend than a matrix.

Figure 6-3. Closed Digraph

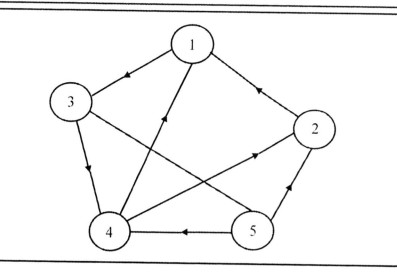

Figure 6-4. Unidirectional Digraph

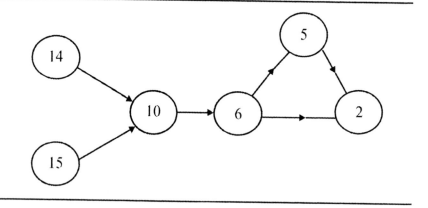

A simple hypothetical matrix is constructed in Figure 6-5 to show how the Boolean process works. Assume that after research, it is found that the main elements of a problem are: (1) Farmers, (2) River, (3) Chemical Processor, (4) Goods Producer, (5) Water Aquifer, and (6) Households. These are arrayed as the rows and columns of the matrix with a 1 in the cells where there are deliveries and with a 0 in the cells where there are no deliveries. The digraph for this matrix is laid out in Figure 6-6 with the deliveries noted on the edges.

Figure 6-5. Simple Social Fabric Matrix

Delivery Component / Receiving Component		Households 1	Water Aquifer 2	Goods Producer 3	Chemical Processor 4	River 5	Farmers 6
Households	1	0	0	0	0	0	0
Water Aquifer	2	1	0	0	0	0	0
Goods Producer	3	1	0	0	0	0	0
Chemical Processor	4	0	1	1	0	0	0
River	5	0	0	0	0	0	0
Farmers	6	0	0	0	1	1	0

The digraph in Figure 6-6 can be used to organize further research and to collect data. Different parts of the system require different kinds of expertise, such as soil scientists, chemists, economists, rural sociologists, water quality engineers, and so forth. Those parts can be assigned and the researchers can decide what kind of research needs to be done to complete the system. They will not see themselves as specialists whose work is disconnected from others. Each researcher will know with whom to coordinate and the kind of information that must be provided to other researchers. The data from the digraph can be stored in a common relational data-management system. Because of the importance of deliveries to a system, the delivery from component to component serves as the columnar headings in such data systems, as

Figure 6-6. Simple Social Fabric Matrix Digraph

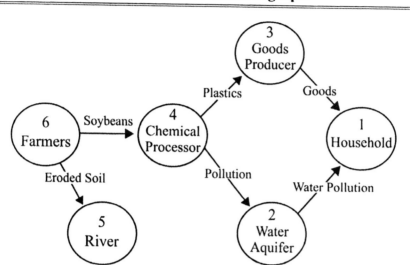

Figure 6-7. Data Management Spread Sheet

Years	Soybeans	Eroded Soils	Plastics	Pollution	Goods	Water & Pollution
	6 ⟶ 4	6 ⟶ 5	4 ⟶ 3	4 ⟶ 2	3 ⟶ 1	2 ⟶ 1

in Figure 6-7. The headings in Figure 6-7 are taken from Figure 6-6.

The relationships among the various deliveries can, through this kind of research, be discovered and built into the digraphs and spreadsheets so that if a policy change is made in one part of the system, the impacts from policies can be identified throughout the system as alternative policy scenarios are developed. For budgeting purposes, it allows the determination of consequences per dollar spent on programs. The columns of the spreadsheet can be added for different programs and those totals compared to the budgets of the different programs.

The digraphs by themselves are very useful for conveying to the research group the structure of the system. In addition, they allow for quick identification of gaps. If the digraph is not fully connected, then it is not fully defined and must be more fully articulated. As the research team makes changes in either the digraph or matrix, a computerized system can be utilized to translate the change to the other researchers. In addition, graph theory has been developed which will allow digraphs to be used as analytical tools themselves. Techniques exist for comparing the capacity of various parts of the system to determine where shortages and surpluses will develop. The surplus may, for example, be excess hazardous waste.

Instead of thinking just in terms of introducing policy changes, other changes such as environmental accidents, terrorist acts, or tornadoes can be introduced into the matrix to determine the direct and indirect deliveries that take place throughout the matrix and digraph.

When it is possible to trace the sequences and linkages in a system, problems and their policy solutions appear much different. For example, there is a global interest in the wetlands located in the state of Nebraska because they are used as a staging area for migrating birds from Russia, Northern Europe, Canada, and Mexico. As farmers have continued to drain the wetlands, the bird population has been severely impacted. The assumption was that farmers convert wetlands to farmland for profit, with the conclusion that the condition of the wetlands was a result of farmers' profit-maximizing mode of thinking. Thus, the policy advice for saving wetlands was a pecuniary solution of offering payments to farmers for not converting the wetlands. That policy was not successful. Analysis found that draining the wetlands was not due to the farmers' profit decisions. When social psychologists were consulted, it was found that an important determinant of the farmers' decisions was their belief in the destruction of wildlife habitat irrespective of whether it was profitable. The belief system in conjunction with the increased availability of drainage technology is responsible for the wetland destruction. The more complete analysis that included all the relevant elements leads policymakers to more relevant policy alternatives.

Comparison to General Systems Principles

The next task will be to show the relation between the SFM methodology and the twelve GSA principles and characteristics explained in Chapter 4 and to explain how the SFM can be used to apply GSA principles. SFM digraphs will be used to explain the extent of the SFM's congruence with GSA principles rather than using matrices. Recall, however, that digraphs do not exist separate from matrices.

System Defined

The SFM approach is consistent with the definition of systems concepts. It defines a whole that transcends the constituents. It contains defined components. It allows for the integration of scientific findings, field observations, and databases and it provides a set of elements together with a definition of the relationships among the elements. The SFM approach encourages the development of the rich and subtle complexity of the system while ordering the complexity for analysis and database development.

Openness

GSA clarifies that a system is open to its environment. The SFM digraph allows investigators to determine if they have constructed such a system. Some component nodes will be from the system and some from the environment outside the system. Figure 6-8 is used to illustrate that the SFM approach is an open-system approach. The matrix and digraph contain the information on the flow of energy, information, and material. In terms of GSA, the system being investigated is the internal system, and the environment outside the system is the external system. If the system of interest in Figure 6-8 is the "Roman" system, the "Arabic" and "English" systems are external because they are outside the system of interest. There would only be interest in the inputs from and outputs to element 8 because that is the entry to the external system. The external environment would be "black-boxed" with no interest in its detail. The only interest in the external system is the impact of the Roman system on the inputs and outputs. If all three systems are in the original matrix in order to give an initial broad understanding, the matrix can be partitioned and disaggregated down to the Roman system. Then the other systems, except for node 8, would be dropped out of the digraph. On the other hand, the study might start with only the Roman system, and the matrix later expanded to include the systems connected through 8.

If we assume that the Roman system is an ecological system, that the Arabic system is a one-directional economic production system concerned about delivering goods to node 4D for distribution to a group of consumers represented by node 1, and that the English system is a

Figure 6-8. Convergence of Balanced, Unidirectional, and Centralized Systems

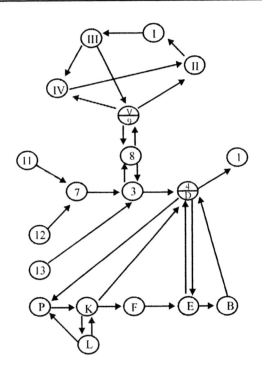

modern governmental system which collects taxes from groups and distributes goods and services to groups; we can begin to understand some current ecological problems. Industry 8 delivers technological processes to the natural environment and extracts natural resources from the ecological system at V/9. In turn, processing industry 3 pays dollars for the extracted resource. It should be noted that there is no information or requirement connection between 8 and I to relate (cause 8 to internalize) to the indirect pollution impact to I, or impact on I's carrying capacity and ability to regenerate. In turn, the decision to buy from 8 by 3, and agreement to a price by 3, does not reflect ecological system impacts.

The government, 4D, is also disconnected from the ecosystem in a direct sense. Its connection is limited to an indirect one of buying-from industry 3. Its payment of a market price thus does not relate to impacts on the ecosystem. The consuming group, 1, to which the government is delivering the product is not made aware of the ecological system. Likewise, taxpayer groups K, E, and B are not making ecological calculations or decisions when they are paying their taxes.

Figure 6-8 allows us to see the potential of the SFM to be utilized to understand and make policy for overlapping regional networks that make up global systems. Wolfram Elsner has explained: "The increasing role of global forms of decision-making, whether they be hierarchical, like multinational firms, or cooperative and networked, has raised the question for the potential of regional resources to be developed through networking. . . . [T]here are specific conditions in which the regional potential for networking is weakened as well as conditions in which it is strengthened, under the regime of globalization."[21] Therefore, for network policymaking and governance in a global system, network building needs to be based on the study of institutions, the changes in institutions that evolve from policy, the impact of that evolution, and the comparison of the alternative networks that will transpire as a result of the implementation of alternative policies. Figure 6-8 can be viewed as three overlapping regional systems. Their refined definition in a SFM context will provide valuable contextual information for the formation of regional and international policies. "Networks exhibit a whole range of characteristics that qualify them as solutions to problems of complexity, uncertainly, repeated direct interdependence, pluriformity and instability in decentralized processes."[22] If those characteristics can be modeled in their real-world context, policy cannot be made to take advantage of potential solutions. The SFM can be used to define the networks as they exist for making social, economic, and ecological system policies.

Nonisomorphic

The SFM conforms to the nonisomorphlic principle which emphasizes that systems are not the isomorphic sum or reflection of the elements.

As illustrated in the example just presented (Figure 6-8), it is clear that the system is not a reflection of the mining industry, and that taxpayers are not aware of ecological entities, nor are all system deliveries the sum of decisions of the human agents. Rather than being isomorphic, systems are transactional with transactions among components and elements being guided by system rules that are across particular organizations and individual elements that are transacting.

The matrix and digraph of the SFM approach also allow for disaggregation into subsystems in a manner that the subsystems can be brought back together into the SFM. Because investigators know the deliveries of the subsystem into the overall matrix before partitioning the matrix, subsystem investigators can structure analysis so that those deliveries continue to be emphasized and are available for relinking the subsystem to the matrix.

Equifinality

With the SFM digraph, it is possible to observe alternative paths through which a system is achieving the same result. This diversity of means leads to system redundancy, which protects the system if one means becomes damaged or if the flow becomes slowed or disrupted.

Several different Boolean algebra manipulations can be performed on the matrix and digraph to help discover alternative restoration scenarios, for example, in the case of natural resource damage. The Boolean manipulations can be used to determine redundancy and transitiveness, and to optimize policies. It is possible to determine, with the use of the Boolean matrices, how many different paths (redundancy) there are in the system to accomplish the same deliveries or sequence of deliveries. The computer request can be made to indicate all like paths or a request can be made to identify all of a particular kind of path sequences.

In addition to observing redundancy, it is possible to use the SFM to generate alternative paths which the investigators may want to consider researching. These are, of course, matrix manipulations. Therefore, a note of caution is in order, such manipulations are completed in order to generate *potential* paths that *might* not otherwise be discovered. The skeleton matrix can be used to generate a Boolean reachability matrix in order to generate all the transitivity paths which do not currently exist in the system. Transitivity is the condition where if element A reaches B with a delivery, and B reaches C, then A's delivery is required in order for C to receive a delivery. A transitive system will demonstrate a chain of paths to fulfill the conditions of transitivity. The potential paths may be utilized in two ways. The first is to determine if there are real-world delivery paths that have been overlooked and the second is to represent potential policy paths that are relevant for building new deliveries.

As with transitivity, optimality paths can also be computer generated with Boolean matrices. Optimality paths indicate the paths to shorten the distance through the digraph network. Again, these paths are generated to determine if there are alternative delivery paths *to be considered*, not necessarily to be implemented.

System Components

The social fabric approach requires the specification of all component elements relevant to the problem. Without this kind of specificity, it is not possible to know the source of deliveries that are being made among component elements.

Control and Regulation

Rather than attempting to determine a system by examining objects and elements alone, the SFM methodology makes control and regulation an integral part of the process. It does this first by laying out antecedents and successors in the delivery process; one delivery perforce leads to another and, therefore, has control over the receiving elements. Second, and more explicitly, real-world control and regulation elements are integrated into the matrix as separate matrix rows and columns, and into the digraph as separate nodes. They include social belief criteria as well as technical and natural criteria. Social beliefs and technological requirements stand behind explicitly stated regulations. Thus, the beliefs need to be integrated and their consequences must be identified, as they are in the matrix and digraph. Likewise, the regulatory organizations which enforce normative criteria are included.

Hierarchy

System hierarchy can be expressed with the SFM digraph. Given a digraph set, for example the English set in Figure 6-8, it is clear that the system is made up of a full set of B, D, E, F, K, L, and P. Any other set within that set will be a subset to the full set, such as the set P, K, and L; and consistently, the set K and L is a subset of P, K, and L. This demonstrates system hierarchy which can be used to determine and specify the deliveries and constraints among the levels of the hierarchy.

Flows, Deliveries, and Sequences

As explained above, the entire SFM approach is based on flows, deliveries, and sequences, as emphasized in GSA.

Negative and Positive Feedback

In some respects, the terms negative and positive feedback are misnomers. To assume a *feedback*, as opposed to a *feed forward*, assumes one part of the system is forward and another part is backward. That is true only in a one-directional growth system. As "feedback" is added, the system ceases to be one-directional and begins to become balanced. The digraphs designed above indicate that feedback loops are only another term for element delivery. If we return to Figure 6-8, we see that the Roman system has the most feedback among the elements, while the Arabic system, as expressed, has no major feedback cycles.

From a SFM analysis of a system, three needs become apparent, as can be observed in Figure 6-8. First, it is apparent that new information deliveries need to be established to, for example, deliver information on ecosystem impacts to producers, buyers, sellers, government agencies, consumers, and taxpayers in the system. For example, in Figure 6-8, the government deliveries to 1, E, and P do not take into consideration the social cost of ecosystem mining. With the matrix we can go cell by cell for each row to determine which decision nodes need to receive new information and what policies are needed for its delivery. A second need which can be observed is the need for new material flows. For example, in Figure 6-8 it is clear that there is no delivery sequence to recycle natural resources back into the ecosystem. Third, a SFM analysis allows us to identify where new regulations, control criteria, and organizations need to be established.

Differentiation and Elaboration

The SFM approach creates a digraph for observing systems and a concomitant database. As the system evolves through differentiation and elaboration, new elements will be added to the matrix and nodes to the digraph. Moreover, new deliveries will be indicated in the relevant matrix cells and corresponding edges in the digraph. By comparing the new matrix, digraph, and database with the original, system evolution can be observed and measured.

By observing the full system, it will be possible to determine whether there has been structural change or just a change in the delivery path due to equifinality. For structural change to occur, it is necessary for the control mechanisms such as beliefs, rules, and requirements to change. These can be monitored and noted throughout the SFM process. With new policies, statutes, and court decisions, new elements are added to the matrix rows and columns. With new technology, new criteria, rules, and requirements are added. As technology changes, the matrix will allow us to see what new criteria (N_T) must be met by what institutions, and thereby to anticipate changes in the social structure and natural environment.

Real Time

The SFM digraph is consistent with activity sequencing called for by real-time systems. Traditional time concepts and clocks are not sufficient for the space-time coordination needed to solve social and ecological problems. It is necessary to be able to model temporal relationships relevant to particular systems. Activity sequencing "puts social problems into a system, which is ideally timed by the succession of events relevant to that system, that is, by *social time*. In other words, reference to universal or clock time becomes secondary to the internalized timing which is defined by the nature of the activity sequence structure."[23]

A SFM digraph can be used to represent the sequence of relations and the direction of deliveries among the components of the social system. Such articulation can be used to plan communication networks, or transportation systems, or pollution controls, or whatever needs to be coordinated in a timely manner. The sequence of events and flows follow one another in an order prescribed by the system, thus making for temporal order. "Frequency of events during a period of time is critical; thus rate is also one of the ways that time impinges on social behavior. For all these elements of social coordination, the term *timing* is useful . . . timing is an intrinsic quality of personal and collective behavior. If the activities have no temporal order, they have no order at all."[24] Temporal order can be articulated with the SFM digraph.

Evaluation

The SFM is used to detail all the entities which contribute to a system. Their contributions are the basis for valuation of a system and its parts. As has been clarified, there is no common denominator which provides one measuring mechanism for a system. The relationships and entities of a system, especially a complex system, call for an array of different kinds of measures and indicators in order to define and evaluate the system. With such measures, it is possible to establish indicators to evaluate a system and policy alternatives. With modern systems, a flood of social indicators come to mind. With regard to agriculture, for example, we might consider exports such as pesticides, sediments, and food contaminants; resource modifications like species diversity and land use patterns; sustainability indicators such as indications on tillage practices and soil organic matter content; contamination indicators like pesticide residues in soil, water, and animals, biomarkers, and heavy metal concentration; and socioeconomic indicators such as farm income and population shifts. Dollar income is included as one of many indicators, but not as *the* measure.

A SFM analysis provides a wealth of information for the evaluation process. Evaluation is about determining what is better and worse, what is improvement, and what is degradation. As systems philosopher Richard Mattessich has stated, "to answer the question of how to improve the system, one needs criteria for and measures of effectiveness."[25] There are a number of socioecological system criteria, control mechanisms, and norms contained in the SFM. *"A system has a goal or purpose either (1) because the inner or mentalistic aspect of the system is developed highly enough so that norms emerge out of this system . . . or (2) because some norms are imposed, in one form or the other, from outside upon the system."*[26]   The SFM documents and shows the importance of both kinds of norms. As social and ecological systems develop, new entities with control properties develop to normalize relations and deliveries in the system; for example, social belief criteria and control mechanisms from the natural environment. In addition, policy control mechanisms are part of the system and included in the SFM description. Mattessich and others have stated that these norms and criteria are the most important system entities in the system. Thus, their condition and ability to guide need to be evaluated. If they are unable to work because of the paucity or abundance of deliveries, they are of less value. The control mechanisms of an oceanic system, for example, may be misfiring because they are overwhelmed with an excess delivery of urban sewage. As another example, evidence indicates that farmers in Iowa have strong belief criteria to protect the groundwater, yet they are polluting it through the use of farm chemicals because their ability to deliver consistent with their beliefs is hampered by the inadequacy of financial, educational, and institutional flows. The condition and welfare of the norms and control mechanisms are important, and their effectiveness can be evaluated through the SFM.

The SFM can be used to determine how effective are the normalization controls by measuring the system flows and deliveries that result from those norms. How great the value of the controls is determined by the degree to which the system is functioning according to a normalized flow. Standard techniques can be used to determine the "goodness of fit" or deviation from the norm.

Evaluation deals with different kinds of concerns for systems such as diversity, stability, transformation, and restoration. Although relevant for all kinds of systems, they are explained here with ecological system concerns. They are:

Diversity Evaluation. There is a concern for biodiversity in ecosystems in terms of the number of species, the inventory of the species, and the redundancy through equifinality. The SFM approach provides information on all three. The species would be a row and column element in the matrix in some cases, and in other cases a cell delivery; for example, a river delivering fish. In order to know either how much a species delivers to another element, or how much is being delivered, it would be necessary to have information on the kind and number of species. Once the basic SFM and digraph are constructed, it is possible to list the species and sum their inventory. It will, therefore, be possible to evaluate ecosystems with regard to biodiversity and to determine whether there are too many or too few of a species, consistent with the carrying capacity of the ecosystem. It will also be possible to determine the degree of equifinality with regard to redundancy, as explained above. If there are more paths for maintaining species, the *system* is more valuable from a biodiversity valuation criterion point of view.

Stability Evaluation. There has been considerable interest in stability evaluation of two kinds. The first is the stability of the system due to vulnerability of the elements. The second is the vulnerability of the system as a whole.

With regard to the first, the SFM can be used to rank the most important relationships and "nerve" centers within a system. By valuing the importance of the centers within the system, system vulnerability can be ascertained. If the system becomes more vulnerable through the destruction of one node over another, then that one is more valuable than the other. The SFM can be used to measure the relative importance of the elements and nodes within a system. For example, the greater the number of 1's in a SFM row, the more deliveries that element is making to other elements. Or, stated differently, the more 1's in the row, the more other elements are dependent on that element. The greater the number of 1's in a column, the more that element is receiving from other elements. Other elements cannot continue to function (process deliveries) if that element cannot continue to receive.

While the greater centricity of a system gives the central node in a system more value, greater centricity makes the system more vulnerable. There is literature to suggest that more diversified ecosystems, for example, are more stable. Following from that, it is possible to compare the stability of systems by comparing their degree of centricity in the SFM digraph. If a system is more centrally organized, it is more vulnerable, and therefore less valuable. If two systems are the same except that one has a few large nodes upon which the system is dependent, then it is less valuable.

Transformation Evaluation. From the SFM database, a normalized flow which must be maintained can be determined, and that normalized value can be used to evaluate alternative economic production projects which are being introduced to transform a socioecological system. No new production project can be introduced without disrupting an ecosystem; thus, some of the normalized flows will have to change. However, by normalizing the flows in the SFM digraph and establishing a spectrum around that norm to establish how far it is safe for the system to deviate, different projects can be judged according to their "goodness of fit." The less the new project causes deviations from the normalized system, the greater its value. It may, of course, be decided that changes can be made in the original ecological system, thereby establishing a new norm.

If a decision is made to change the system flows from the original, the same evaluation procedure can be followed with a new normalized delivery level. The SFM database can be used to indicate the impacts of the new flow levels throughout the system. For each delivery upon which an economic project will impact, whether quantitative or qualitative, the normal delivery needs to be established and the project's deviation from it determined. If the project falls within the critical threshold, it is acceptable. If it best fits the overall flow levels, it is the most efficient. Every project has a multitude of impacts, and they should be considered in a systems approach to minimize transformation costs.

Restoration Evaluation. The establishment of restoration costs to restore a damaged ecosystem, for example, is not a case of evaluation. It is an operational action to convert the damages into a budget sufficient for restoration in order to increase the value of the ecosystem. The July 1989 ruling in *State of Ohio v. U.S. Dept. of the Interior* on this subject is consistent with this view. The Court stated that "restoration is the proper remedy for injury to property where measurement of damages by some other method will fail to compensate fully for the injury."[27]

Restoration costs are not necessarily market costs in the sense that the prices to be paid for the equipment, labor, and materials are not necessarily established by a competitive private market system. Some prices are explicitly impacted through governmental subsidies and taxes, and others are charges by other governmental agencies to do the cleanup. In addition, many of the private sector prices are determined shelf prices to get the job done. A SFM digraph model of an ecosystem would be helpful in tracing the indirect impacts of a toxic or hazardous substance spill to help trace how the spill is delivered through the system, and therefore all the costs which must be undertaken for restoration.

Restoration evaluation is different than restoration cost. The valuation aspects of system restoration can be completed with the SFM. First is the selection of the optimal restoration alternative. Restoration projects themselves can also change an environment. Thus, they should be judged as outlined above in the section on "Transformation Evaluation." The optimal alternative is the one which generates flows to return the ecosystem to its original purpose and structure without creating other adverse deliveries outside the threshold level for the system. The second valuation aspect of restoration is to minimize the use of resources in the cleanup. As explained above, the SFM offers digraphs to illustrate alternative paths that exist to accomplish the same purpose and maintain its capacity. Therefore, if one path is damaged and there is a redundancy of equifinality paths, it may be that the ecosystem will be able to fulfill its goal without utilizing as many resources. Some of those paths may appear feasible and viable, and therefore can be tested against other alternatives (as explained above) to determine if they are more valuable restoration alternatives.

Summary

The SFM approach to policy research emphasizes (1) that humans and their governments want to solve problems and (2) that every system should be analyzed in conjunction with its environment. Thus, the SFM was developed, consistent with the principles of GSA, to explain how the flows and delivers among system components are creating a problem within real-world systems. The methodological approach forces the researcher to assume a holistic systems-oriented view of the process under investigation. The SFM can be used to more precisely define systems so that data bases can be created consistent with the SFM description, normative gaps and inconsistencies can be identified, and research agendas can be refined and assigned to determine the way that normative criteria and institutions should be structured through policy for a more instrumentally instituted process.

The usefulness of the SFM for policy analysts can be summarized as follows:

- The SFM is a policy research and information tool that is designed to encourage analysts and policymakers to ask relevant questions and to compile information and knowledge to answer those questions.

- The SFM defines a whole that transcends the components and describes their integration. Matrix and digraph manipulations enable researchers to order complex systems and develop integrated databases to understand problems.

- The SFM calls for the integration of diverse kinds of data and the establishment of cumulative databases for monitoring policy impacts and responding to system changes.

- The SFM requires the researcher to consider a system as an open system because relevant deliveries to and from the natural environment and other social systems are to be included.

- The SFM and the digraph focus upon the whole and the interrelationships of the parts, rather than aggregating the parts. Both also enable disaggregation for the analysis of subsystems.

- Multiple and alternative paths to the same system results can be observed and understood from the SFM and digraph.

- The SFM requires specification of relevant values, beliefs, institutions, attitudes, technologies, and ecological components as needed for the problem under consideration.

- Criteria from social beliefs (N_B), technology (N_T), and the natural environment (N_E), which are mechanisms for control and regulation, are included as separate matrix rows and columns. Consequences of such criteria can therefore be identified, measured, and acted upon with policy.

- The SFM and digraph are based upon sequenced flows and deliveries.

- Component flows can be identified to serve as variables for statistical analysis.

- Feedback loops can be understood and uncovered through the process of constructing the SFM.

- New elements can be added to the SFM and digraph, as differentiation and elaboration occur in order to model and evaluate evolutionary change.

- Within the SFM and digraph, clock time is secondary to the timing of sequenced flows of the system. The ordering and timing of these flows are illustrated in the SFM and digraph for understanding current problems and for long-range planning.

- To determine the most efficient approach to solve a problem, a different SFM and digraph can be completed for each alternative policy or program in order to compare the consequences of policy alternatives.

CHAPTER 7

ILLUSTRATIONS OF THE
SOCIAL FABRIC MATRIX

For purposes of illustration, four pared down examples of social fabric matrix (SFM) studies are presented in this chapter. The SFM and digraph approach to system analysis has been used to study diverse contexts in different nations. Included are studies on the European telecommunications industry, the impact of United States farm policy on the agricultural community and on the global flow of commodities, interlocking corporate directorships, the Australian bee pollination industry, a low-level radioactive waste compact system, the AID's system in the United States, and African wildlife planning. The SFM, in conjunction with systems dynamic techniques, has also been utilized as a planning tool between confrontational groups consisting of farmers and environmentalists in Australia. Two major SFM studies that are currently ongoing are a forestry policy analysis in the northwestern United States and a socioecological study being completed in Thailand. This chapter is devoted to abridged presentations of SFM and digraph studies in order to illustrate systems derived from the SFM approach and the diverse kinds of problem and policy contexts to which it can be applied. The first two presented also demonstrate the use of the computer program *ithink*,[1] which is a program that can be useful for processing data consistent with the SFM.

Contracts and Costs in a Corporate/ Government System Network

The SFM and digraphs for this study were completed in order to determine how social beliefs codified in contracts influence costs and to complete a 30-year projection of costs for a waste disposal system.[2] Thurman Arnold explained in his *Folklore of Capitalism* that modern industrial systems are the integration and coordination of diverse kinds of huge organizations. Corporations, government agencies, universities, and multi-state compacts, for example, are the kinds of organizations that make up the organizations in this case study. They function together, are dependent upon each other, and their content and function are determined by the actions taken in the contracted networks that integrate the organizations. The policy-oriented purpose of the SFM analysis is to analyze the ramifications of a cost-plus contract arrangement that is very influential in determining the costs and activities of a particular corporate/government network.

The network included the five states of the Central Interstate Low-Level Radioactive Waste Compact (CIC) to include the CIC's policymaking Commission, American Ecology Corporation (AEC) which was selected as the developer of a nuclear waste facility, a number of major corporations, and several government agencies. The broad network of institutional organizations is found in the SFM in Figure 7-1 and its digraph in Figure 7-2. The figures emphasize the pre-operational phase of the development process with only one entry (row and column 17) serving as the subsequent operational phase. A more refined elaboration of the cost aspects of both phases is presented below. The institutional component that created the compact system is the U.S. Congress (row and column 1) by passing legislation for states to create a system of compacts for the management or disposal of low-level radioactive waste. To guide the system's creation and function, Congress directly delivered criteria, rules, regulations, authority, and responsibilities to federal government agencies and to the five CIC states. These federal agencies, in turn, promulgated rules, regulations, requests, and inspections that were delivered to the five states [cell (2,4)], to the State of Nebraska which was selected as the state for a disposal site [cell (2,5)], and to American Ecology Corporation (AEC) which is the corporation selected to develop the radioactive waste facility [cell (2,13)] for the CIC.

On the left and across the bottom of the SFM digraph in Figure 7-2 are arrayed the government agencies. However, the most powerful organizations in the network are the major global corporations placed at the top of the digraph. They are major energy corporations and Bechtel. The energy corporations are the major generators of radioactive waste in the five-state region and the source of debt financing for the CIC to use to pay AEC which in turn pays Bechtel for development

Figure 7-1. General Social Fabric Matrix of CIC System

Delivering Components \ Receiving Components		U.S. Congress (1)	Federal Govt. Agencies (2)	Non-Host State Govts. (3)	Nebraska State Govt. (4)	Nebraska State Agencies (5)	Local Monitoring Committee (6)	Boyd county Communities (7)	Central Interstate Compact (8)	Major Generators (9)	Minor Generators (10)	Contract (11)	Agreement (12)	American Ecology Corp. (13)	Bechtel (14)	Other subcontractors (15)	Sub-subcontractors (16)	Operational Phase (17)
U.S. Congress	1	1	1															
Federal Govt. Agencies	2			1	1			1						1				1
Non-Host State Govts.	3							1										
Nebraska State Govt.	4					1		1										
Nebraska State Agencies	5						1							1				1
Local Monitoring Committee	6																	
Boyd county Communities	7																	
Central Interstate Compact	8							1		1		1	1	1				1
Major Generators	9									1				1				1
Minor Generators	10																	1
Contract	11							1	1					1	1			
Agreement	12							1	1					1				
American Ecology Corp.	13					1		1				1			1	1		1
Bechtel	14													1		1		
Other subcontractors	15													1		1		
Sub-subcontractors	16														1			
Operational Phase	17							1						1				

Source: Hayden and Bolduc 2000.

activities. The major generators and Bechtel became the most powerful organizations by controlling the contracts and agreements.

Since the problem of interest is with cost, an in-depth analysis of the contracts is needed because that is where the normative criteria that determine costs are specified and from which enforcement of criteria are made. The contracts[3] were thoroughly analyzed to discover the major criteria, requirements, and institutional organizations relevant for projecting costs in the subsequent operational phase. These contractual entities are entered into the SFM in Figure 7-3 to explain what the components deliver to each other for the problem area.

The social belief criteria are expressed in Figure 7-3 (rows 1 through 9) as the designated sections in the contract from which the criteria gain their legitimacy. Important to note is the process through

Figure 7-2. General Social Fabric Digraph Network of CIC System

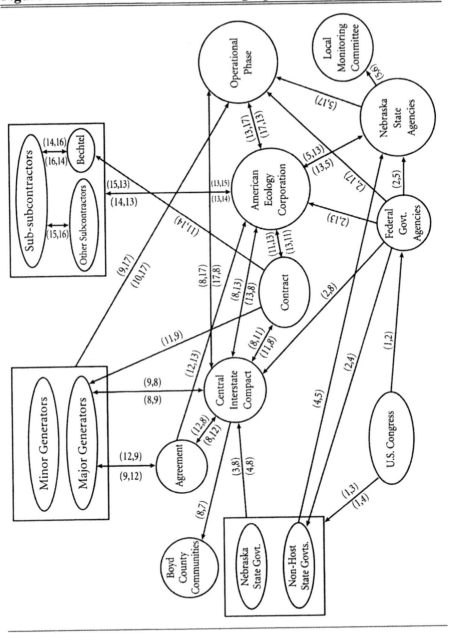

Source: Hayden and Bolduc 2000.

Figure 7-3. Social Fabric Matrix of Contracts and Costs during CIC Pre-Operational Phase

		Receiving Components →	A. Sub-subcontractor without adder	B. Sub-subcontractor with adder	C. Bechtel's operating expenses less payroll & property	D. Bechtel's home-office payroll	E. Bechtel's non-home office payroll	F. Bechtel's equipment costs	Bechtel National, Inc.	American Ecology Corporation (AEC)	Central Interstate Compact (CIC)	Major Generators	Operational Phase
	Delivering Components ↓		10	11	12	13	14	15	16	17	18	19	20
Criteria and Rules	Contract 4.01 (a)(i)(A)(B)(D) & (E)(1) & (2)	1	1										
	Contract 4.01 (a)(i)(C)(D) & (E)(1)(2)	2		1									
	Contract 4.01 (a)(i) & Technical Service Agreement II & III	3			1	1	1	1	1	1	1		
	Contract 3.05, 3.07 & 6.01	4								1	1		
	Contract 3.02 (b), 4.01(b), 4.02, & Amend. 3 (4) & (9) (6.01)	5								1	1		
	Agreement & Contract 2.01(a)(b),3.02,3.03,4.01 & Amend. 3	6									1		
	Agreement & Contract Amend. 3 (2)	7											1
	Agreement & Contract 2.01 & 4.02	8										1	1
	Agreement & Contract 2.01 & 6.01	9									1		
Institutional Components	A. Sub-subcontractor without adder	10							1				
	B. Sub-subcontractor with adder	11							1				
	C. Bechtel's operating expenses less payroll & equipment	12							1				
	D. Bechtel's home-office payroll	13							1				
	E. Bechtel's non-home office payroll	14							1				
	F. Bechtel's equipment costs	15							1				
	Bechtel National, Inc.	16	1	1					1				
	American Ecology Corporation (AEC)	17							1		1		1
	Central Interstate Compact (CIC)	18								1		1	1
	Major Generators	19									1		
	Operational Phase	20											

Institutional Components

Source: Hayden and Bolduc 2000.

which belief criteria are selected and enforced in a society. The belief criteria included in the contract and actually enforced in society are inconsistent with those suggested in the Code of Federal Regulations 10.61.[4] The belief criteria outlined through the democratic process are not the ones implemented in this case.

Each cell where there is a delivery is designated with a 1. By going cell by cell in Figure 7-3 to each cell with a 1, the cost formula defining costs is derived. The SFM provides a means to describe the general context, to define connections among components in the context, and to convert cellular information to mathematical expression where appropriate. Symbols A through F (see rows 10 through 15 in Figure 7-3 and corresponding nodes in Figure 7-4) designate both the category of the contractor and the invoice expenses of the same contractor. Some of the cell deliveries in Figure 7-3 are explained here in order to illustrate the aggregating process of contracts as found in the SFM. They are:

Cell (1,10) Contract's Delivery to A: The delivery of interest in cell (1,10) for defining costs is the coefficient delivered to sub-subcontractor A from the Contract. That coefficient is the coefficient that may be used to multiply costs before sending the invoice to Bechtel for reimbursement. A is the expenses of a sub-subcontractor whose contract with Bechtel does not allow for a cost-plus adder; thus, the coefficient delivered in cell (1,10) is 1.

Cell (2,11) Contract's Delivery to B: The Contract terms in row 2 of the SFM are delivered to B. B is the expenses of a sub-subcontractor who may add 10 percent to its cost before sending its invoice to Bechtel. Thus, the coefficient delivered in cell (2,11) is (1 + .10).

Cell (3,12) Contract and Technical Service Agreement's Delivery to C: The CIC/AEC Contract and the Technical Service Agreement between AEC and Bechtel join together to determine the relevant delivery in this cell. C is Bechtel's own operating expenses, exclusive of payroll expenses and capital equipment costs that are delivered internally to Bechtel. The contract delivers a coefficient of (1 + .10); thus, the C expenses on the invoice can be increased by 10 percent.

Cell (3,13) Contract and Technical Service Agreement's Delivery to D: D is Bechtel's home-office payroll expenses for the project. Due to contractual deliveries, they are increased by about 41.5 percent, and, in turn, 78 percent of that total is added for a coefficient of (1+ .415)(1 + .78).

Cell (3,14) Contract and Technical Service Agreement's Delivery to E: E is Bechtel's non-home-office payroll expenses

Figure 7-4. Social Fabric Digraph Network of Contracts and Costs during CIC Pre-Operational Phase

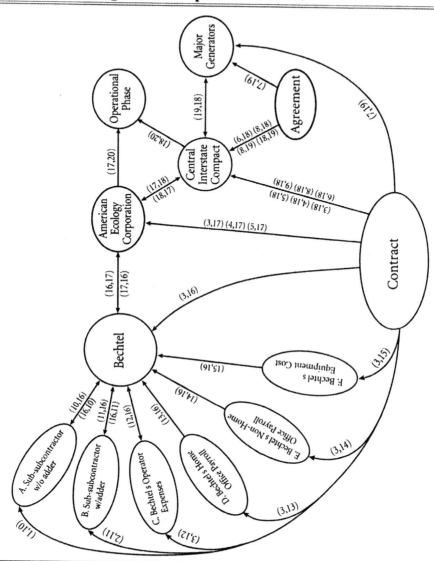

Source: Hayden and Bolduc 2000

for wages and salaries. The contractual delivery is $(1 + .415)(1 + .425)$, whereby E is increased by 41.5 percent, and that total is increased by 42.5 percent.
Cell (3,15) Contract and Technical Services Agreement's Delivery to F: The coefficient delivered in Cell (3,15) is one which is to be applied to F. F is capital equipment that is bought by the subcontractor, Bechtel, who selects the equipment and owns it.
Cells (10,16) and (11,16) Sub-subcontractors' Deliveries to Bechtel: The sub-subcontractors' expenses, multiplied by the contractual coefficients delivered to them (as explained above), are delivered as invoices to Bechtel. They are, therefore, A and $B(1 + .10)$ when delivered.
Cells (12,16), (13,16), (14,16), and (15,16) Bechtel's Internal Cost Categories Deliveries to Bechtel: Bechtel's own expenses are delivered to Bechtel, with their coefficients as defined above. They are: $C(1 + .10)$, $[D(1 + .415)][1 + .78]$, $[E(1 + .415)][1 + .425]$, and F.
Cell (3,16) Contract and Technical Service Agreement's Delivery to Bechtel: Cell (3,16) provides the contractual authority for Bechtel to apply the coefficient $(1 + .05)$ to invoices it receives from sub-subcontractors A and B and $(1 + .10)$ to its own internal expenses as they are bundled and presented with their own internal coefficients, as presented in the cellular explanation just preceding this one.
Cell (3,17) Contract and Technical Service Agreement's Delivery to AEC: Cell (3,17) delivers to AEC the coefficients to apply to invoices received from subcontractors such as Bechtel. The contractual authority is provided for AEC to multiply subcontractors' invoices by $(1 + .05)$ and to multiply its own expenses by $(1 + .08)$.
Cell (16,17) Bechtel Delivery to AEC: Given the authority explained above to utilize the contractually defined cost-plus coefficients, Bechtel sends an invoice to AEC as follows: $\{A + [B(1 + .10)]\}\{1 + .05\} + [C(1 + .10)] + \{[D(1 + .415)][1 + .78] + [E(1 + .415)][1 + 425]\}\{1+.10\} F(1 + .10)$.
The aggregating character of the Contract becomes clear with some very large cost-plus adders being authorized, often without a rational economic explanation. D, for example, is increased by 41.5 percent, and that total increases by another 78

percent, as was explained above. That total, with adders, is increased again by 10 percent before sending the invoice to AEC. E is treated similarly. F is an equipment gift to Bechtel, ultimately paid for by the CIC. Bechtel selects the equipment, gets to own it, and in addition adds 10 percent onto the cost of the gift.

Cell (3,18) Contract and Technical Services Agreement Delivery to the CIC: Cell (3,18) establishes the authority for the cost-plus invoices to be submitted to the CIC for reimbursement.

Cell (17,18) AEC Delivery to the CIC: With regard to the Bechtel part of the cost-plus process, AEC takes the invoices sent to it by Bechtel, adds another 5 percent of that total, as provided in cell (3,16), and sends the total invoice to the CIC for reimbursement. Thus, the Bechtel part of the formula becomes:

$$\{\{A + [B(1 + .10)]\}\{1 + .05\}\}\{1 + .05\} + [C(1 + .10)][1 + .05] + \{\{[D(1 + .415)][1 + .78] + [E(1 + .415)][(1 + .425]\}\{1 + .10\}\}\{1 + .05\} + [F(1 + .10)][1 + .05].$$

As clarified in the discussion of cell (3,16), AEC may also add a 5 percent adder to the costs of other subcontractors as well as add an 8 percent adder to AEC's own expenses. Therefore, the total invoice, PV, sent to the CIC is as found in formula (1).

$$PV = \{\{A + [B(1 + .10)]\}\{1 + .05\}\}\{1 + .05\} \qquad (1)$$
$$+ [C(1 + .10)][1 + .05] + \{\{[D(1 + .415)][1 + .78]$$
$$+ [E(1 + .415)][1 + .425]\}\{1 + .10\}\}\{1 + .05\}$$
$$+ [F(1 + .10)][1 + .05] + G + [H(1 + a)] + DEQ + [I(1 +$$
$$.08)] + \{[J(1 + .365)][1 + .75] + [K(1 + .365)][1 +$$
$$.425]\}\{1 + .08\} + [L(1 + .08)] + M + N$$

PV stands for the present value of the invoices sent to the CIC. In addition to the Bechtel part of the formula contained in Figure . . .[7.3], all contractors, subcontractors, and sub-subcontractors are included (symbols G through N are defined in Table . . . [7.1]). When the formula is solved, the percentage coefficient for each term is as found in Table . . . [7-1].[5]

Examples from Table 7-1 can be used to indicate why the costs of the project became so exorbitant and why cost overruns were common. For example, sub-subcontractors are not allowed a cost-plus adder, yet their costs, A, invoiced to the CIC are 110 percent of costs because Bechtel and AEC are allowed to include their own percentage adders. Bechtel's home-office payroll, D, is invoiced at 291 percent of cost, its non-home-office payroll by 233 percent, and so forth. The information in formula (1) and Table 7-1 also clarifies that all incentives in the Contract are to increase cost functions without disincentives or penalties for increasing cost functions. The Bechtel part of formula (1) has the highest cost adders, and it is also the part that has received the most payments from the CIC. As an example, $1,000,000 of direct costs becomes a budget of $2,269,850.63 when all adders, summing to $1,269,850.63, are added as defined by formula (1).

The SFM provides a means to describe the general context, to articulate particular socioeconomic components embedded in that context, to define connections among the components, and to convert cellular information to mathematical expressions where appropriate. The analysis completed above allows us to see how the cost-plus formula builds through the Contract to create a financial burden to be carried forward to be amortized and paid during the operational phase. The total delivered to the operational phase is about one-half billion dollars. It includes the major generator financing ($325 million) as explained above, AEC financing contribution ($20.7 million) as explained above, construction loan ($91.6 million), financial assurance loan ($41.5 million), and AEC's subsidiaries' interest ($12.6 million). The analysis above pinpoints the particular provisions in the Contract that need to be renegotiated to reduce the financial burden.

The aggregating character of the cell deliveries becomes clear. This demonstrates large cost-plus adders in the final aggregated formula. For example, in the final analysis of the whole SFM, Bechtel charges 291 percent of its home-office payroll, 233 percent of its non-home-office payroll, and 116 percent of Bechtel's capital equipment that was already paid for by the CIC. Similarly, AEC charges 258 percent of its non-home office payroll. The total costs are delivered to the subsequent operational phase.

The SFM for the operational phase is found in Figure 7-5. Components 1 through 5 are debt components from the pre-operation phase that are to be paid during the 30 years the facility is to operate. Components 6 through 23 are the organizations and processes integrated for the operations phase. The SFM of Figure 7-5 is reported,

Table 7-1. Cost-Plus Percentages that Multiply CIC Costs

Percentage	Definition of Terms
PV=110% of A	A is cost of sub-subcontractors without cost-plus contract
+121% of B	B is costs of sub-subcontractor with cost-plus contract adder equal to 10%
+116% of C	C is Bechtel's operating expenses, exclusive of payroll & capital costs
+291% of D	D is Bechtel's home office payroll
+233% of E	E is Bechtel's non-home office payroll
+116% of F	F is Bechtel's capital equipment bought by CIC
+G	G is stipulated sum or unit cost contract awarded by Nebraska DEQ
+(1 + a) of H	H is cost-plus contracts awarded by Nebraska DEQ
+DEQ	DEQ is in-house costs of Nebraska DEQ
+105% of I	I is stipulated sum or unit-cost contract awarded by DEQ
+108% of J	J is AEC's operating expenses exclusive of payroll and equipment costs
+258% of K	K is AEC's home office payroll
+210% of L	L is AEC's non-home office payroll
+108% of M	M is AEC capital equipment bought by CIC
+N	N is lobbying and extraordinary legal expenses of AEC

Source: Hayden and Bolduc 2000.

Figure 7-5. Social Fabric Matrix of Institutional Components during CIC Operational Phase

		1	2	3	4	5	6	7	8	9	10	11	12	13	14	15	16	17	18	19	20	21	22	23
Major Generators Financing	1						1																	
Construction Debt	2							1																
Financial Assurance Debt	3								1															
AEC Financing	4									1														
AEC General Interest	5										1													
Major Generators Amortization	6																			1				
Construction Amortization	7																			1				
Financial Assurance Amortization	8																			1				
AEC Financing Amortization	9																			1				
General Interest Amortization	10																			1				
Depreciation Schedule Application	11																			1				
Rate Setting Process	12																						1	
CIC Debt Fund	13																						1	
CIC Administration	14																							
Community Improvement Account	15																1							
Boyd County Agencies	16																							
Nebraska State Agencies	17																1							
American Ecology Corporation	18												1											
AEC Charges & Payment Fund	19													1	1	1	1		1				1	1
AEC Waste Storage	20																							
Waste Generators	21																		1	1				
Major Generators	22																							
Subcontractors	23																							

Source: Hayden and Bolduc 2000.

modeled, and analyzed with the assistance of the computer program, *ithink*. The SFM digraph (Figure 7-6) is in the format of the *ithink* program. The modeling capabilities of *ithink* can be used consistent with the theories and process concepts of the SFM.

The analysis begins by presenting the highest level framework and proceeds to an explanation and analysis of the deliveries among particular sectors as found in the SFM in Figure 7-5. The high level digraph, found in Figure 7-6, presents the main institutional organizations. The heavy lines in Figure 7-6 indicate the general flows among sectors; the thinner lines represent deliveries of rules, regulations, or criteria. As the high level digraph in Figure 7-6 is elaborated with information from the SFM of Figure 7-5, the more complex digraph of

Figure 7-6. High Level Mapping of CIC Network Structure

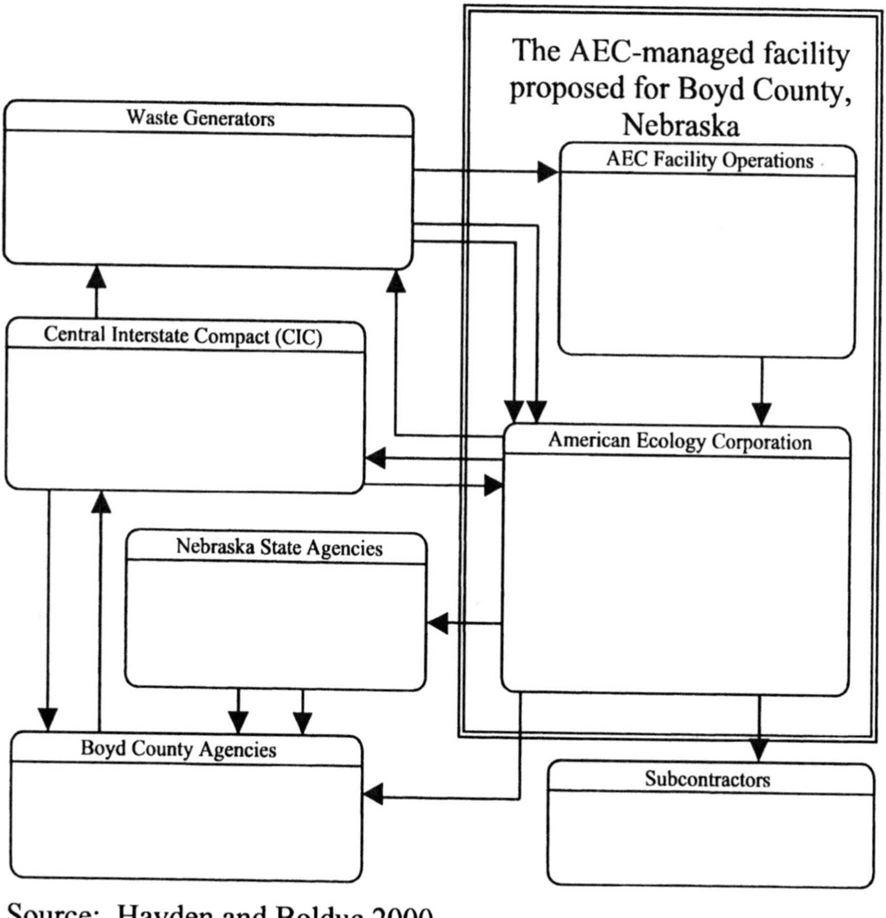

Source: Hayden and Bolduc 2000.

Figure 7-7 is derived. It is presented with *ithink* symbols (that are also utilized in Figure 7-11) defined as follows:

ithink components

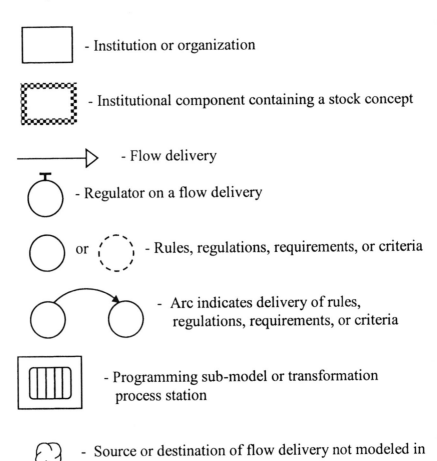

- Institution or organization

- Institutional component containing a stock concept

- Flow delivery

- Regulator on a flow delivery

or - Rules, regulations, requirements, or criteria

- Arc indicates delivery of rules, regulations, requirements, or criteria

- Programming sub-model or transformation process station

- Source or destination of flow delivery not modeled in the system

The analytical capabilities of *ithink* are used to complete calculations on a year-by-year basis for the 30 years of waste facility operation. From the analysis, it was found that the charges that would be necessary to cover the contractual costs are greater than $18,500 per cubic foot of radioactive waste for all the years waste is to be received during operation of the facility. This means the policy alternative of building a new waste facility under the current contract is not a viable alternative. This is because there is currently excess capacity for low-level radioactive waste in the private sector and the charge by Envirocare of

Utah is about $90 per cubic foot for the lower spectrum of low-level radioactive waste, while the charge by Chem-Nuclear Systems of South Carolina is about $470 per cubic foot for the full spectrum of low-level radio active waste.

This SFM study presents a new way to analyze legal contracts that guide systems; a way that allows for connections to be made between contract provisions and the consequent flows and deliveries. This analysis also demonstrates that social beliefs are not vague abstractions, rather they are criteria embedded in rules, regulations, and requirements as expressed and enforced in contractual obligations. Moreover, beliefs are divided and developed among a whole host of institutions and obligations. Policy analysts cannot determine the efficacy of a policy or program until normative criteria are identified and their consequences measured.

The SFM analysis of this corporate/government network confirms and magnifies the finding Henry Maine made late in the 1800s when he found that the base of society had evolved from status to contract. The contractual element has continued to grow in importance and the battles over contractual form have a great influence on the working of the modern business, industrial, and social processes.

An Analysis of the Daily Federal Fund's Market

The purpose of the second study presented is to provide a framework to analyze the daily federal funds market that is crucial to Federal Reserve policy.[6] In terms of problem orientation, the primary problem the study is attempting to remedy is the lack of a current description of the daily federal funds market that integrates several important aspects of the market's day-to-day functioning. This is an application of the SFM in a macroeconomic setting. First, a SFM is completed that includes only beliefs and institutions that are defined broadly. The SFM is found in Figure 7-8 with cells where there are deliveries from the row components to the column components indicated with a 1 in the cell. All the cells with a 1 are explained fully in the original study, however, only brief explanations are provided here for some of the cells. They are:

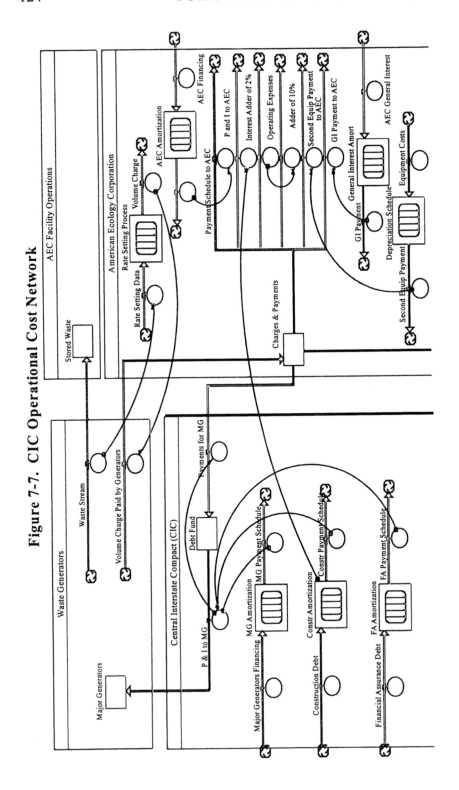

Figure 7-7. CIC Operational Cost Network

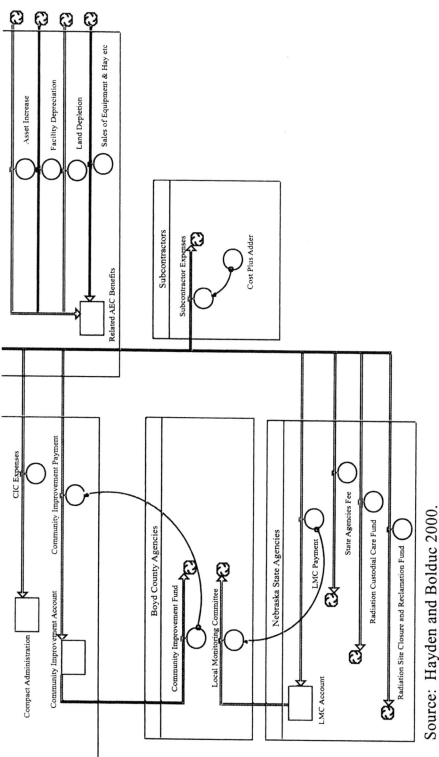

Source: Hayden and Bolduc 2000.

Figure 7-8. Social Fabric Matrix of the Daily Federal Funds Market

Delivering Components \ Receiving Components		Beliefs		Institutions							
		Legal Tender Laws	Federal Reserve Act	US Courts/Legal System	Federal Reserve Banks	Board of Governors	FOMC	Banks	Open Market Desk	Treasury	Primary Dealers	Congress
		1	2	3	4	5	6	7	8	9	10	11
Legal Tender Laws	1			1	1			1		1		
Federal Reserve Act	2			1	1	1	1	1	1			
US Courts/Legal System	3											
Federal Reserve Banks	4					1	1		1			
Board of Governors	5				1		1	1				
FOMC	6							1				
Banks	7						1			1	1	
Open Market Desk	8							1		1		
Treasury	9				1			1	1		1	
Primary Dealers	10							1	1	1		
Congress	11	1	1	1	1	1	1	1		1	1	

Source: Fullwiler 2001.

Cells (4,9) and (9,4) Federal Reserve Banks Delivery to the Treasury and Treasury delivery to Federal Reserve Banks: "Federal Reserve Banks, as established in the Federal Reserve Act, are fiscal agents and depositaries for the Treasury. The Treasury's accounts are maintained at the Federal Reserve Banks, from and to which all payments to and from the Treasury are made. At the end of each business day, the Treasury's accounts are combined into one account at the Federal Reserve Bank of New York."[7]

Cell (5,6) Board of Governors Delivery to the Fed's Open Market Committee (FOMC): "As established in the Federal Reserve Act Section 12A, members of the Board of Governors are members of the FOMC."[8]

Cells (5,4) and (5,7) Board of Governors Delivery to Federal Reserve Banks and to Banks: "The Board of Governors authorizes changes in the discount rate and sets reserve requirements for banks on demand deposits."[9]

Cell (6,8) FOMC Delivery to the Open Market Desk: "The FOMC delivers instructions in its directive after each FOMC meeting to the Desk. Since 1988, the directive has set the target for the federal funds rate that the Desk attempts to maintain on average on a daily basis. The FOMC's annual Authorization for Domestic Open Market Operations is the source of the Desk's authority for purchasing different types, quantities, and maturities of financial assets in its open market operations."[10]

Cell (7,7) Banks Delivery to Other Banks: "In the federal funds market, as explained above, banks borrow and lend reserve balances held in accounts at Federal Reserve Banks. Banks generally borrow reserve balances for the purpose of meeting reserve requirements or in order to clear payments."[11]

Cells (7,9) and (9,7) Banks Delivery to the Treasury and Treasury Delivery to Banks: "Thousands of commercial banks nation-wide are depositaries of the Treasury in the Treasury Tax and Loan (hereafter, TT&L) account system. The Treasury attempts to maintain a balance of $5 billion each day ($7 billion during certain periods when tax collection flows are largest). When net balances in the Treasury's account are above (below) this amount, the Treasury adds to (calls from) balances held by TT&L participating banks. Tax payments (and Federal Government spending) occur as simultaneous debits from (credits to) a taxpayer's bank account and the taxpayer's bank's reserve account, and credits to (debits from) the Treasury's account at the Federal Reserve."[12]

Cells (7,10) and (10,7) Banks Delivery to Primary Dealers and Primary Dealers Delivery to Banks: "Commercial banks hold deposit accounts for primary dealers, which are debited and credited when the Desk engages in open market sales or purchases of securities. Open market purchases (sales) therefore result in credits (debits) to reserve accounts held by banks at the Federal Reserve."[13]

Cells (8,9) and (9,8) Open Market Desk Delivery to the Treasury and Treasury Delivery to the Open Market Desk: "Each business morning, the Desk and the Treasury have a conference call to share information regarding expectations of both the Desk and the Treasury of flows into and out of the Treasury's account on that day. The discussion is part of the Desk's preparation for the day's open market operations. Net changes to the Treasury's account below $300 million are not normally offset by the Treasury through calls from or adds to the TT&L accounts; rather, such a small expected reduction (addition) to the Treasury's account results in the same sized addition (reduction) from the Desk's planned operations."[14]

Cells (8,10) and (10,8) Open Market Desk Delivery to Primary Dealers and Primary Dealers Delivery to the Open Market Desk "The Desk purchases securities from and sells securities to primary dealers in the secondary market in its daily open market operations. Primary dealers also provide the Desk with market information to help guide the Desk's implementation of monetary policy. Primary Dealers are selected by the Desk as counter parties for open market transactions and are required to participate in both open market operations and Treasury actions."[15]

Cell (11,1) and (11,2) Congress Delivery to Legal Tender Laws and to the Federal Reserve Act: "Legal Tender laws are written and have been periodically amended by Congress, a power given to Congress in the United States Constitution."[16]

Cell (11,3) Congress Delivery to the US Courts/Legal System: "Laws written by Congress are interpreted by the courts/legal system in the settlement of disputes."[17]

Figure 7-9 is a digraph picture of the SFM in Figure 7-8. The nodes in Figure 7-9 are the components in Figure 7-8 and the directed edges (lines) indicate the deliveries among the components. Edges with one arrow represent one cell with the cell number printed on the edge. Edges with two arrows indicate deliveries in both directions between two cells. Figures 7-8 and 7-9 indicate the functions that need to be defined for the relationships among the components, the indicators that need to be identified consistent with the functions, and the databases that need to be accumulated of the indicators.

The SFM was reduced to be concerned about only part of Figures 7-8 and 7-9 and to define each in a more refined way with more cells. Thus, in terms of rows and columns, the new SFM (not shown) has more rows and columns than Figure 7-8. The new SFM is expressed in digraph format in Figures 7-10, 7-11 and, 7-12. The digraphs are expressed in the format and symbols of the computed program *ithink* (defined above).

Figure 7-10 is an overall picture of the system and provides the broad outline of the main institutions and their connections. A more

Figure 7-9. Social Fabric Digraph of the Daily Federal Funds Market

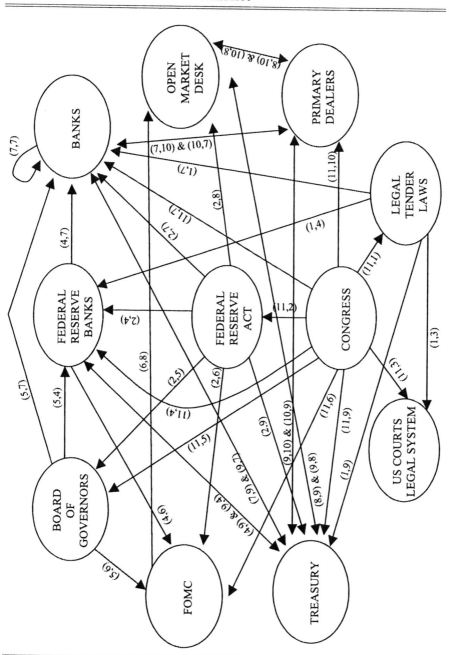

Source: Fullwiler 2001.

detailed digraph of the Open Market Desk part of the system in Figure 7-10 is illustrated in Figure 7-11. Figure 7-11 is presented to demonstrate the complexity and relationships of the parts of the Open Market Desk.

The application of the SFM approach allows for numerous findings that are important for Federal Reserve policymaking. Some are:

- The complexity of the system clarifies that simple models of Federal Reserve policy ignore many potential system reactions to policy.

Figure 7-10. High Level Mapping of the Daily Federal Funds Market

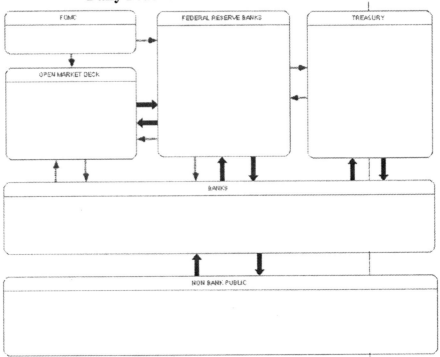

Source: Fullwiler 2001.

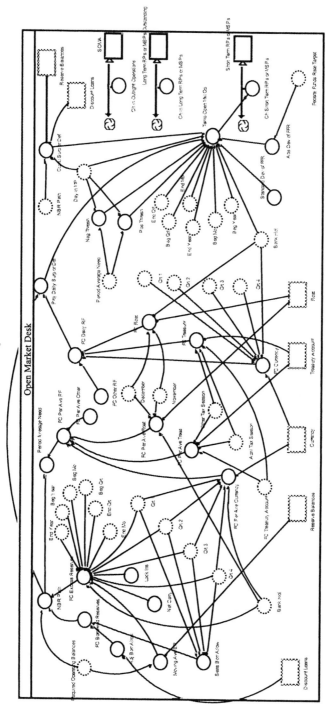

Figure 7-11. Digraph of the Open Market Desk System

Source: Fullwiler 2001.

Figure 7-12. Digraph of the Federal Reserve System

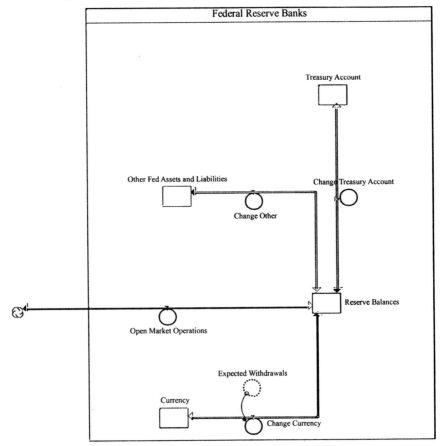

Source: Fullwiler 2001.

- "Perhaps most importantly, the demand for reserve balances and the supply of reserve balances cannot be considered independent either for theoretical or statistical modeling purposes, because they are intimately related through feedback loops."[18]
- A proper measure of total reserve balances, for the purpose of understanding monetary policy, includes the quantity of required reserves.   This is particularly the case given that nearly all (if not all) reserve balances are currently held primarily for meeting payment-clearing needs."[19]

A Socioeconomic Analysis of Rural Poverty and Livelihood Strategies in a Village in India

This third SFM is based on field studies and surveys conducted in Theethandapattu, a dryland village in the state of Tamil Nadu of Southern India.[20] It concentrates on modeling the social, ecological, and technological components that focus on the impacts of seasonality on livelihood strategies. Livelihood strategies are the strategies relied upon by rural families to mitigate various risks and impacts they confront such as loss of income and reduced entitlements and capabilities. The study is oriented toward the impacts confronted due to seasonality and the livelihood strategies practiced in response to seasonal cycles. The purpose was to identify and understand the complex interwoven web of causation among the myriad economic, social, cultural, and ecological factors that transact as a system to cause and deepen poverty. A focus on livelihood strategies guides the modeling of the complex system in which activities are contained and by which families attempt to make a living.

The complexity of the system is illustrated in Figure 7-13. Figure 7-13 is a SFM of components that make deliveries to provide the system of livelihood strategies. The deliveries among some of the most important components of the SFM are laid out in digraph format in Figure 7-14.

The SFM approach revealed a number of conclusions, to include that:

- Vulnerabilities are created by the working of the complex system. Linkages were identified between components that interact in the system to create deprivation.

- Livelihood strategies (in response to falling incomes and loss of capabilities) of subsistence farmers, wage laborers, and large farmers have triggered a simultaneous decrease in both the demand for and supply of labor at the village level.

- The character and pattern of livelihood strategies differ by class, caste, and rural group because of different roles and positions with respect to the system's social, economic, ecological, and technological components. Through the identification of the deliveries made by farmers, it was found that lack of food in the region is because most production is the growing of non-food commercial crops for urban and global markets. The poorest farmers deliver food to their own families because they are

Figure 7-13. Social Fabric Matrix of Components that Generate Livelihood Strategies in Theethandapattu, Tamil Nadu, India

	Receiving Components \ Delivering Components	36 Forest	35 Water	34 Aquifers	33 Land	32 Seasonality (Climate)	31 Media (TVs, radios. etc)	30 Farm Implements	29 Tractors	28 Draft Power	27 Electric Pumps	26 Wells	25 Political System	24 Laws	23 Infrastructure	22 Hospitals, Health Care	21 Schools	20 Caste	19 Harvest Festival	18 Birth & Death Ceremonies	17 Marriage	16 Markets	15 Family	14 IMF & World Bank	13 Government	12 Livelihood Strategies	11 National Ag Banks	10 Mechanized Farming	9 Govt. Jobs=Job Sec	8 Social Sanction	7 Role of Women	6 Material Well-Being	5 Ceremonies & Festivals	4 Income Security	3 Economic Growth	2 Astronomy & Astrology	1 Supernatural
		Ecological					Technology						Social Institutions															Social Beliefs								Value	
Value	1 Supernatural																													1			1				
	2 Astronomy & Astrology																													1							1
Social Beliefs	3 Economic Growth												1	1	1	1	1					1		1	1	1	1					1		1			
	4 Income Security													1			1				1	1	1	1	1	1	1	1	1			1			1	1	
	5 Ceremonies & Festivals																	1	1	1	1					1				1		1				1	1
	6 Material Well-Being													1			1				1	1	1	1	1	1							1	1	1	1	1
	7 Role of Women													1			1	1		1	1	1	1	1	1	1				1				1	1		1
	8 Social Sanction																	1		1	1		1						1		1		1				1
	9 Govt. Jobs= Job Sec													1			1	1			1		1		1	1			1	1				1	1		
	10 Mechanized Farming												1	1	1							1		1	1	1	1				1			1	1		
Social Institutions	11 National Ag Banks	1	1	1	1	1	1	1	1	1	1	1	1	1								1		1	1	1		1						1	1		
	12 Livelihood Strategies	1	1	1	1	1	1	1	1	1	1	1	1	1	1	1	1	1	1	1	1	1	1	1	1	1	1	1	1		1	1	1	1	1		
	13 Government	1	1		1		1						1	1	1	1	1					1		1	1	1	1	1	1			1		1	1		
	14 IMF & World Bank		1		1	1			1	1	1	1	1	1	1	1	1					1		1	1	1	1	1				1		1	1		
	15 Family		1		1	1				1				1	1		1	1		1	1	1			1	1			1	1	1	1	1	1	1		
	16 Markets	1	1		1	1		1	1	1	1			1	1							1		1	1	1	1	1			1	1		1	1		
	17 Marriage				1	1	1	1		1				1			1	1		1			1			1			1	1	1		1		1		
	18 Birth & Death Ceremonies																	1			1		1			1				1			1				
	19 Harvest Festival					1				1																1							1				
	20 Caste	1	1		1						1	1	1	1						1	1		1		1	1				1	1		1				
	21 Schools					1							1	1	1								1		1	1					1				1		
	22 Hospitals, Health Care												1	1	1							1			1										1		
	23 Infrastructure	1	1		1		1		1				1	1		1	1					1		1	1		1	1							1		
	24 Laws	1	1		1								1		1	1	1					1			1										1		
	25 Political System	1	1		1		1					1		1	1	1	1	1				1			1			1							1		
Technology	26 Wells										1			1								1		1		1	1										
	27 Electric Pumps								1						1							1		1		1	1										
	28 Draft Power							1	1	1												1				1	1										
	29 Tractors									1					1							1		1		1	1										
	30 Farm Implements								1	1												1				1	1										
	31 Media (TVs, radios, etc)												1	1	1		1	1				1			1	1	1										
Ecological	32 Seasonality (Climate)														1							1															
	33 Land													1	1			1			1	1	1	1	1	1	1										
	34 Aquifers														1							1				1	1										
	35 Water													1	1			1				1		1	1	1	1										
	36 Forest													1	1			1				1			1	1	1										

Source: Natarajan 2001.

Figure 7-14. SFM Digraph among Key Components of Livelihood Strategies

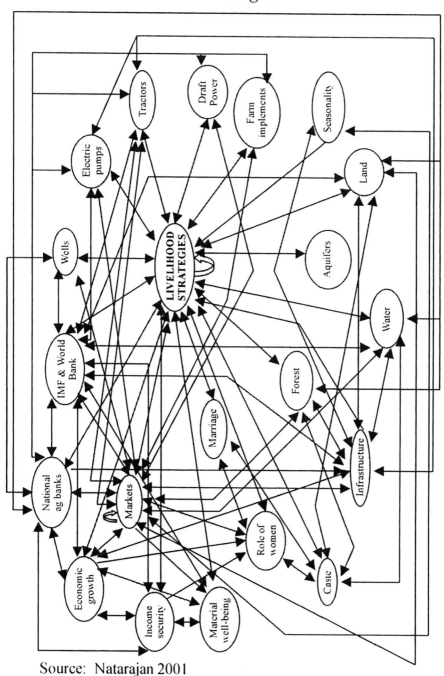

Source: Natarajan 2001

- primarily concerned with meeting their irreducible minimal human needs of food consumption. The wealthier farmers who hold most of the land are more concerned with producing non-food commercial crops for the exterior market.
- Caste is very important in determining how various groups are allowed to respond to seasonality in order to maintain a livelihood. It was found that it is not possible to separate who is poor from why they are poor in a socioecological process. The Indian caste system has long been recognized as important in the religious and social systems, but the SFM allowed for its extensive influence and linkages through the economic, technological, and ecological processes to be discovered and documented. Anyone wanting to deal with agriculture policy in rural India would want to know the working of the institutions and the sundry paths that make for the network of deliveries. The SFM and corresponding digraph allow for tracing the impacts of changing a delivery or set of deliveries.

Assessment of Institutional Performance for Surface Water Management of the Platte River in Nebraska

The final abridged presentation of a SFM analysis is of the policy network for the management of the surface water of the Platte River in Nebraska.[21] This SFM and digraph clarify the extensiveness and complexity of policy networks established for economic and ecological systems and clarify how crucial it is for policymakers to understand how changes in rules and regulations generate changes in the network.

The SFM framework is utilized to analyze how water resource allocation systems are constructed, implemented, and operated by ecological and social variables that continue to evolve. The framework provides a systemic understanding of the structure, function, and performance of the surface water systems by the regulatory authority of the Platte River. The performance of the institutional authority is evaluated in terms of ecological consequences as the test of instrumental efficiency. The enforcement of criteria and governing rules requires the assistance of administrative and regulatory authorities that are structured as hierarchical organizations that respond to the ongoing process of water resource management.

The administrative and regulatory structure has evolved in response to changes in the volume and quality of the water in the Platte. The development of industry, especially irrigation for the agriculture industry, caused a reduction of average peak flow into the Platte River of 86 percent between 1902 and 1970. The water carrying channel decreased by about 70 percent resulting in a 72 percent destruction in lowland grasslands and wet meadows. Non-point sources of pollution increased due to agricultural chemicals such as fertilizers and pesticides. The reduction of quantity and quality has consequently endangered species of the Platte. This includes birds (such as whooping cranes, bald eagles, least terns, and piping plovers), and aquatic insects that provide the main ingredient in the food chain of fish and other aquatics. Further, the destruction of the riverine system of the Platte has damaged recreational activities and industries connected to hunting, fishing, and bird watching. The regulation and management network for the Platte has evolved in order to allocate the surface water flow among competing interests.

The general SFM model and concepts used to guide the research are displayed in the SFM digraph found in Figure 7-15. It is consistent with the conceptual relationships between social beliefs and institutions, and among institutions as displayed above in Figure 2-2 and explained in Chapter 6. To demonstrate evolution of the system, different matrices were constructed for different legislative and judicial eras. The matrices were for each era following the passage of (1) the 1885 Riparian Rule, (2) the 1902 Federal Reclamation Act and the 1930s New Deal Policy, (3) the five Nebraska Environmental Laws of the 1970s, and (4) the 1984 Nebraska Legislative Bill 1106 which is the current era.[22] Consistent with social theory regarding differentiation, each SFM for each succeeding era becomes more complex.

The current SFM for the latest judicial and legislative era is found in Figure 7-16. Instead of using a 1 in each cell where there is a delivery, codes for actual deliveries are used in Figure 7-16. As examples, "a" is used in all the cells where Legislative Bill 1106 delivers legal criteria to the institutional organizations numbered 2 through 10 and 12, and "b" is used in all the cells where the institutional organizations deliver policy criteria to other institutional organizations. An elaboration of some of the cells is as follows:

Figure 7-15.   General Social Fabric Matrix for Surface Water Management in Nebraska

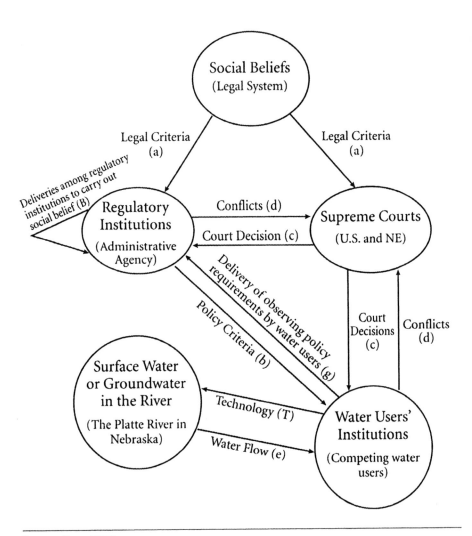

Source: Yang 1996.

Cells (1,6) and (1,12): These cells indicate that the principle of water resource management in Nebraska is rooted in the 1895 Prior Appropriation rule which emphasizes "first in time and first in right" and is recognized as the basic rule of water management.[23]
Cells (2,3), (2,3), (2,4), (2,5), (2,6), (2,7), (2,8), (2,9), (2,10), and (2,12):   These cells indicate the delivery of legal criteria for the

Figure 7-16. Social Fabric Matrix for the Current Era

Criteria / Delivering Components	#	1 Prior Appropriation Rule	2 Legislative Bill 1106	3 Army Corps of Engineers	4 U.S. Fish and Wildlife Service	5 Game and Parks Commission	6 Dept. of Water Resources	7 Natural Resources Districts	8 Water Management Board	9 Director of Natural Resources	10 Natural Resource Commission	11 NE Government (Governor)	12 Nebraska Supreme Court	13 Public Irrigation Districts	14 Agricultural Users	15 Public and Private Power Dist.	16 Domestic Users	17 Municipal Users	18 Instream Users I (GPC)	19 Instream Users II (NRDs)	20 Surface Water in the Platte	21 Groundwater	22 Endangered Species
Prior Appropriation Rule	1						a					a											
Legislative Bill 1106	2			a	a		a	a	a	a	a	a							a				
Army Corps of Engineers	3				B										d	b							
U.S. Fish and Wildlife Service	4			B									d										
Game and Parks Commission	5						B	B	B					d	b	b	b		b	b			
Dept. of Water Resources	6					B		B	B			B	d	b	b	b	b	b	b	b			
Natural Resources Districts	7					B	B							d		b		b	b	b			
Water Management Board	8					B	B								b	b	b		b	b			
Director of Natural Resources	9							B															
Natural Resource Commission	10							B															
NE Government (Governor)	11							B	B														
Nebraska Supreme Court	12		c	c	c	c	c							c		c			c	c			
Public Irrigation Districts	13					g	g		g				d										
Agricultural Users	14					g		g	g	g													
Public and Private Power Dist.	15					g	g		g				d										
Domestic Users	16					g	g																
Municipal Users	17					g	g																
Instream Users I (GPC)	18					g	g	g	g			d											
Instream Users II (NRDs)	19					g	g	g	g			d											
Surface Water in the Platte	20													e	e	e			e	e		e	e
Groundwater	21													e	e		e	e			e		
Endangered Species	22														e	e		e	e		e		

a - Legal Criteria.   b - Policy Criteria.   B - Delivery of institutuional consultation among regulatory institutions.
c - Conflict of interests.   d - Decisions of the court.   e - Water flow.   g - Response of water users' institutions.

Source: Yang 1996.

resolution of conflict delivered by the 1984 Legislative Bill 1106 to regulatory agencies.[24]

Cells (3,4), (4,3), (5,6), (5,7), (5,8), (6,5), (6,7), (6,8), (6,11), (7,5), (7,6), (8,5), (8,6), (9,8), (10,8), (11,8), (11,9), and (17,16): These cells represent the communication or interlocks among regulatory institutions under Legislative Bill 1106.   The digraph outlining the

relationships among these cells is Figure 7-17.   Each cell and its
corresponding edge in Figure 7-17 include the deliveries of consulta-
tions, opinions, permits, and directives.[25]

**Figure 7-17. Digraph of Relationships among Regulatory Institutions
under the 1984 Legislative Bill 1106 in Nebraska**

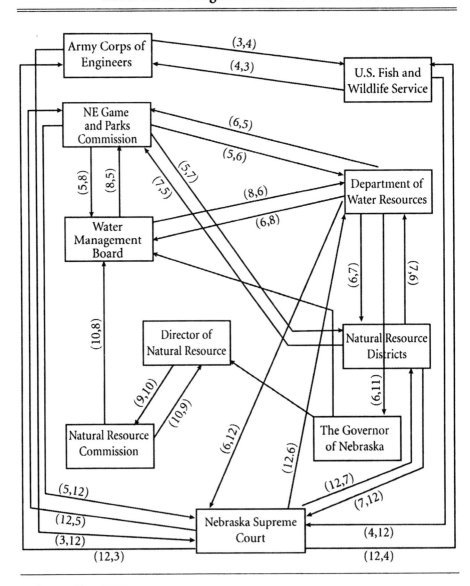

Source: Yang 1996.

The SFM of Figure 7-16 is converted into two different digraphs to illustrate two different concerns. The digraph in Figure 7-17 is concerned with the relationships and deliveries among regulatory organizations, as specified in Nebraska Legislative Bill 1106. The system requires water users to obtain use rights through an application process among the organizations. The digraph in Figure 7-18 is the application process taken from the SFM in Figure 7-16. The SFM and digraphs emphasize (1) the necessity for all parts of the whole to be acting consistent with the whole system in order for the whole system to function effectively and (2) that crucial to the ecological system is the institutional system that maintains and sustains the ecological system.

The final section of the study (1) applied Boolean algebra techniques to the SFM to assess the degree of connectivity and reachability of the matrix and (2) assessed the success of the system in terms of the consequences produced by the system.

Concluding Remarks

The SFM examples presented in this chapter are conclusions taken from abridged versions of the original studies and the more elaborated originals would need to be consulted to obtain the detail necessary for duplicating their approach in other similar settings. The versions presented here:

- Demonstrate applications of the SFM in real-world settings.
- Provide illustrations of the use of the SFM to capture different kinds of complex open systems.
- Reiterate how different problems lead to the study of different contexts.
- Demonstrate that it is unlikely that policymakers can successfully formulate policies and programs without an understanding of the complexity of the system for which policies and programs are being formulated.
- Demonstrate that it is unlikely that policy analysts can select the appropriate program alternative for implementation without a model through which the direct and indirect deliveries of the program can be traced and measured.
- Reiterate that for measurement to be helpful for policy formulation, it must be completed with social indicators that are consis-

Figure 7-18. Relationships among Regulatory Institutions and Water Users' Institutional Organization in the Application Process

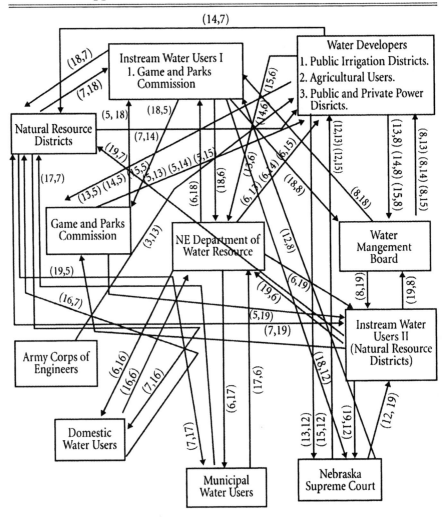

Source: Yang 1996

tent with the context of the problem model. The SFM helps define and provide guidance for selecting and developing appropriate indicators.

- Reinforce and demonstrate the ideas of many system principles such as openness, complexity, hierarchy, positive and negative feedback, and so forth.

- Demonstrate that social beliefs are not vague abstractions but are explicit criteria that are refined, embedded, and enforced by an array of institutions and organizations throughout the network.

- Identify the belief criteria actually implemented and enforced in society as opposed to those wished for or discovered by surveys of the populace.   Thus, beliefs actually functioning to guide a system are not necessarily determined by democratic processes. Examples are the belief criteria in the government/corporate contracts case of the low-level radioactive waste compact and the caste beliefs in the India case.

CHAPTER 8

TIMELINESS AS THE APPROPRIATE CONCEPT OF TIME

In explaining the differences among seven different capitalist systems, the authors of the book, *The Seven Cultures of Capitalism*, emphasize the importance of different concepts of time among the different societies that contain different kinds of capitalist systems.[1] They explained that the different temporal views influence corporate planning, decision making, and production. In a similar manner, we need to be cognizant of different time dimensions for social planning and policymaking. To complete analysis for social planning and policymaking, timeliness is the appropriate concept of time and is consistent with the utilization of the social fabric matrix (SFM).

Important to any policymaking process is the question of *when* actions and events are to occur. To effect new social structures, actions and events must be properly sequenced. An analytical core of sequencing events is time analysis. Early in the Twentieth Century, John R. Commons stated that in addition to knowing what to do, those who have the power and responsibility for planning "must know *what, when, how much* and *how far to* do it at a particular time and place in the flow of events. This we designate the principle of timeliness."[2] This is not consistent with the common approach to time analysis. The more common approach leaves us at the mercy of the concept of passing time.

An exercise the author has performed on numerous occasions is to ask a group of university sophomores to take out a piece of paper and write a brief answer to the following question: What is time? Only once has a student referred to what had been learned in a physics course. Again and again the students express the common bias of Western culture regarding time. "Time is eternally passing." "Time is flowing past almost unnoticed." As one sophomore so vividly stated, "the second that is here now will soon be going over the mountains and travel on to California." Most people are not so graphic in their explanation of passing time, but most believe the common misconception that time is a flowing reality that is constantly passing. The common view is wrong. We have not seen time flowing by, no one has measured it flowing by, heard it, or even documented any telltale signs of its passing, yet the Western mind continues to hold to the common view. The flowing stream of time is so vividly believed that people seriously consider that the loss of aircraft over the Bermuda Triangle could be because of warps in the time stream, and neoclassical economists seriously recommend that education should be delivered to children on the basis of the outcome of a discounted time stream.

Time cannot be seen or touched. It is not a thing at all, but an experience of the human mind. Time is not a natural phenomenon; rather, it is a societal construct. What exist in society are duration clocks and coordination clocks selected by organizations for the sequencing and coordination of events as scheduled by the societal patterns of institutions. Time is the duration of motion between events, and it has no direction. Time measurement is a relative concept that relates one motion to the duration of another motion—a clock.

In modern paradigms, time does not flow, or go, or run from here to there in either a cyclical or a linear pattern. It also does not bridge a spatial gap between points. Neither is it a natural phenomenon. Time is a relative concept that relates one motion to the duration of another motion—a clock. In traditional concepts, motion and succession were thought to take place in time. Today, motion defines time. The duration between successive events is measured by still another motion, such as atomic vibrations or hour clocks or color changes. The successive events are not connected by any rope of time or temporal plane. With so many items in motion, the universe is full of clock candidates. Since these items are moving in a multitude of directions, we might want to say that time is multidirectional. But that would be inaccurate. Time is the duration of motion. It has no direction.

There is no future into which the world is destined to flow except in the sense of the "after" that follows the "before." This certainly removes a good bit of the glow of destiny, inevitability, and evolutionary progress contained in the old concepts of time and future. Or, stated differently, the modern concept of time allows us to use clocks to regulate events rather than ourselves being regulated by the running clock of passing time. This expands the discretionary role and responsibility of modern planners.

Social and Cultural Time Constructs

To find social time, it is necessary to observe the cultural symbols and institutions through which human experience is constructed. Our own Indo-European language imposes the concept of time on us at a very early age, so it is difficult to identify with the concept of timelessness. Societies do exist without a consciousness of time, but most societies possess some concept of time. Conceptualizations of time are as varied as cultures, but most can be categorized by one of three images. Time in the broad sense is viewed as a line (linear), a wheel (cyclical), or as a pendulum (alternating phenomenon). But within these central images, numerous other distinctions are made. One is tense. Is there a past, present, future, or all three? If more than one, to which is the society oriented and in what way? Another is continuity. Is time continuous or discontinuous? Does the continuity or hiatus have regularity? Another is progressiveness. Is evolutionary transformation expected with the passage of time? Still another is use. Is time used for measuring duration or for punctuality? In addition, the metaphysical distinctions are numerous. Is there a mode for measuring time? Is it reversible or irreversible? Subjective or objective, or both? Unidirectional? Rectilinear?

Various Cultural Times

These distinctions come into focus when one studies various cultures. The Pawnee, for example, have no past in a temporal sense; they instead have a timeless storehouse of tradition, not a historical record. To them, life has a rhythm but not a progression.[3] To the Hopi, time is a dynamic process without past, present, or future. Instead, time is divided vertically between subjective and objective time. Although Indo-European languages are laden with tensed verbs and temporal adjectives indicating past, present, and future, the Hopi have no such verbs, adjectives, or any other similar linguistic device.[4] The Trobriander is forever in the present.[5] For the Trobriander and the Tiv, time is not continuous throughout the day. Advanced methods of calculating sun positions exist for morning and evening, but time does not exist for the remainder of the day.[6] For the Balinese, time is conceived in a punctual rather than a durational sense. The Balinese calendar is marked off not by even duration intervals, but, rather, by self-sufficient periods that indicate coincidence with a period of life. Their descriptive calendar indicates the kind of time, rather than what time it is.[7] The Maya had one of the most complicated systems of time yet discovered. Their time divisions were regarded as burdens carried by relays of divine carriers—some benevolent, some malevolent. They would succeed each other, and it was very important to determine who was currently carrying in order to know whether it was a good time or a bad time.[8] (Was this an early explanation of business cycles?) The most common explanations of time among human societies, however, are the cyclical ones.

Cyclical Time

Two examples of cyclical time are those of the ancient Greek and Indian. Socrates believed his teaching was not new—he had repeated it in cycle after cycle for many lives. Aristotle denied the logical possibility of movement in a straight line to infinity and believed that the only everlasting continuous motion is circular motion. An example of cyclical time still in existence is found in India. As in most Indian thought, spirituality plays a major role. For the Western mind to absorb the meaning of Indian time is a difficult task. One is dealing with more than one cycle, with more than one God, with numerous qualities and quantities of time, and with astronomical figures. For example, one day and one night for Brahma, in which a universe arrives and is destroyed, is equal to 4,320,000,000 human years, and this pattern will exist for 100 years of Brahma. But this is only the beginning; there are time cycles within the great cycle, with the *Mahayuga* consisting of four ages, each of which is longer than our historical records. One thousand *Mahayugas* make up a *Kalpa* and fourteen *Kalpas* make up a *Manvantra*, and on and on and on.[9] For the present purpose, it is enough to note that man and history become quite insignificant in comparison to time. This has a tremendous impact on social and economic life. At birth, each person receives a *dharma* which is a moral code and duty appropriate to a given status and pattern of life.

> The joint belief in station and association obligation (*dharma*) and in the related cycle of existence (*karma*) serves as cement in uniting the caste system, and *jajmani* economic system, and all superior/inferior relations. . . .

The idea of *Karma* is additionally associated with a cyclical view of human and natural history. Annual crop and seasonal changes are circularly recurrent. . . . A man's life also is a cycle, and *karma* emphasizes the rhythm of life and death and rebirth. Because the world is viewed in terms of recurring cycles, the notion that a society might advance systematically in any way (i.e., the concept of "progress") does not naturally present itself.[10]

Clearly, time concepts in India affect the economy, as do time concepts in all societies.

Western Time: Traditional

As stated above, Western temporal constructs are consistent with the idea of a linear, flowing stream. "But in the broad sweep of human existence, that way of constructing the dimension of time is relatively rare."[11]   The traditional Western linear time stream came into full bloom with Christianity, although seeds of it had been alive for some time. For example, in the third century B.C., Strato said, "day and night, a month and a year are not time or parts of time, but they are light and darkness and the revolution of the moon and sun. Time, however, is a quality in which all these are contained."[12]   This quote illustrates the appearance of flowing time to explain succession. Through the experience of succession, the appearance of time as a separate eternal entity suggests itself. The reality exists only in the appearance.

What was left of cyclical time among the Romans died with Christ. For Christians, Christ was not a mythical entity. He was on earth and a part of history. Since he died as a man and for all, he could not die again as in cyclical time. Time must be linear. His death "was a unique event. It had happened, once, here on earth, in a certain place at a certain time, and any suggestion that it could happen again was a whisper from the devil."[13]

Augustine rejected Aristotle's connection of time and motion, and avoided any confusion of temporal with spatial concepts. He argued that time was found in the soul and in the mind. Christ died for our sins and man now moved forward into a sacred, prophecy-fulfilling future.[14]   History would flow from one prophesied interval to the next until the final judgment. Christianity became an onward religion. "The time series is linear . . . in the sense that it had a definite beginning at the creation of the world; it will have a definite end when Christ returns at the Last Day; and between these two limits it manifests, at least in a general way, the working out of a consistent Divine Plan. . . . The movement towards its fulfillment does give a direction and purpose to the time series."[15]   Time gave Christians opportunity. "The certitude that time is useful and opportune clearly belongs to the conscious tradition of Christians; for them time is not an inert thing nor is its course a simple chronological unfolding without aim."[16]

The place of the individual Christian was, first, to order life down to the last detail with a view of growing in the love of God, and, second, to forever be alert to accept God's grace.[17]   Thus, the idea of the atomistic individual concerned with growth and progress along the inevitable stream of time becomes apparent. In order to build God's community for His glorification on earth and the saving of souls, a positive work ethic was necessary. One was to be judged not only for good works but for good work as well. A Benedictine monastery's appreciation of labor is revealing: "Labor was to be man's greatest joy and the instrument of his union with God. Industry was the key to the up-building of the new world that Benedict created . . . . Everything was seen as an aspect of work."[18]

Since so much building, caring, and praying needed to be done, with no idea how much time was left, time became precious to the medieval mind. Both the individual and the Kingdom could be called before all the preparation was completed. "Thus, early Christian ideology also created a fertile ground for technological innovation and its employment for earthly improvements."[19]   The more that could be done with passing time, the better. Since individual and community were committed to sacred and earthly accomplishments, accountability and measurement became important. The Christian Middle Ages achieved "an ordered society, sustained by a common and not ignoble belief, in which the individual was not lost but discovered himself, not suppressing but releasing joyous energies, reaching to the sky in the shape of cathedral towers, for the glory of God."[20]

Although the object of the glory has changed considerably, Christian time remains. It is a passing time without qualitative changes, to be used efficiently. At one time the object of the efficiency may be commercial interests, at another technology, at another Mammon, but the concept of time has not much changed.

The Enlightenment really did not alter traditional time concepts. Newton's time was Christian time, and despite Leibniz's argument that time is relative and that instants apart from things are nothing, the traditional concept of time was predominant in the scientific community until much later. Not until Einstein dealt with time on micro and macro levels did the scientific community change. However, the traditional view remains predominant today as the "common sense" view of time, and as the neoclassical economics view. Neoclassical assumptions regarding time are a continuation of their religious predecessors.

In neoclassical thought, individual consumption replaces religious glorification; thus the present becomes more important and pressure exists for individuals to get their fill before their time runs out. This is part of the transition from "biblical time" to "merchant time."

Neoclassical Time-Stream Discounting

Neoclassical thought puts its temporal construct into operation through the use of the geometric mean discounting formula expressed in its simple form as:

$$P_v = \frac{B_1}{(1+i)} + \frac{B_2}{(1+i)^2} \cdots + \frac{B_n}{(1+i)^n}$$

Where: P_v – Present Value
   B – Benefits
   i – interest or discount rate
   n – the nth or final year of the planning horizon

Consistent with its tradition, neoclassicalism borrowed the discounting tool from the world of capitalist finance. It is appropriate for the financier's portfolio analysis, but it is not appropriate as an evaluative tool for social policy. A review of the basic premises and assumptions involved in using the discounting technique makes it clear that they are not consistent with modern science. They are:

Isolation. Time analysis should be built around an isolated extrapolation or projection of segregated benefit and cost flows as opposed to an analysis of process integration. This stems from the analysis being atomistic and isomorphic, instead of holistic.

Importance. Time discounting should occupy a position of major importance in social evaluation. If resources constrain an analysis, then the research and organization of data for discounting should be the last to be eliminated. Thus, enormous research effort has gone into determining the correct discount rate for water projects such as the construction of large dams, for example, with very little research focusing on matters such as the effect of water projects on the quality of community and ecological life, and on the destruction of water.

Measurability. What is important is measurable. Matters that are represented by quantitative measures are the most important. Qualitative measures are ignored.

Consumer Desire. Evaluation of merits and flaws, and benefits and costs should be consistent with a philosophy that promotes consumer desire, even in those cases where there is no vehicle for registering such consumer desire.

Valuing Benefits and Costs. Both benefits and costs can be valued by the same indicator that is a pecuniary common denominator. Pecuniary prices are assumed to serve as a common denominator that takes account of relative scarcities and surpluses.

Closed Future. The future is closed, which assumes that there is a future out there, instead of assuming that there are numerous potential futures.

Big is Beautiful. More is better than less. The passage of time requires progress. The greater the present worth, the more superior a project is judged to be. The present worth is increased by increasing net benefits.

Existence of Time Preference. Neoclassicalists assume, tautologically, that since interest rates accompany saving and consumption by individuals, it follows that individuals have independent time preferences.

Time-Preference Judgment. Individuals have an inherent, intuitive, internal ability to judge time, both durational and continuous. Individuals have independent time preferences and can make judgments regarding them.

The Present is Most Important. The present gets top priority. Even though the additional hospital beds, for example, will not be needed for five years, they should be discounted *through* time and assigned a value for the present period when they are not needed.

Linear Proportionality. In addition, the correct fit for the time preference function is the linear proportionality of the geometric mean formula. Thus, time-stream discounting also assumes that social time preferences are expressed best through percentage rates.

The Commonality of Clock Time.   The same time-measurement scale (clock) has the same meaning in different systems and across different institutions or can be used as a common denominator of different systems. This assumption means it is not necessary to develop a clock consistent with the system.

A Common Discount Rate.   The same discount rate is correct for costs as well as benefits, and is correct for all kinds of costs and benefits.

The Existence of Time.   Time exists as part of the physical world. Therefore, discounting can take place over time and *through* time, and events take place in time. Humans, as well, are in time.

Commodity Value.   Because time exists, it can be scarce. Because it is scarce, it can and should be valued as a commodity.

Rectilinear and Forward Flowing.   Time in neoclassical theory is what usually is identified as the Christian or Indo-European conceptualization of time.   It is one-directional, moving into the future in a linear fashion. Time was switched on in the beginning and is still running.

Unidimensional and Absolute Character.   Time is one-dimensional in its nature, texture, rate, structure, setting, and the like. These characteristics remain absolute in the neoclassical analysis.

Flowing Stream.   Time is a flowing stream; thus the nomenclature "time stream analysis." Like so much of the rest of neoclassical theory, time, as a stream, comes from Newton.   In the beginning of the *Principia*, he said, "Absolute, true and mathematical time, of itself, and from its own nature, flows equally without relation to anything external."   Although this statement is consistent with traditional Christian thought starting at least as early as Augustine, once Newton articulated it, it reified time and ascribed to it the function of flowing.   Although from the beginning this reification was challenged and denied by others of equal fame such as Leibniz, Kant, Einstein, and Planck, they never were able to remove it from the public mind because of the general cultural and religious reinforcement.   Later generations of physicists, astronomers, and philosophers were able to erase traces of Newtonian time from their models, but it is still basic to time analysis in neoclassical economics.

Continuously Passing Time.   Not only is it a stream, but the stream is continuously passing—running on and on and on. Flowing. Flowing. As the moralist cautions, "Make the most of time.   Use it before it goes."

Infinite. Time is an infinite open set. The stream without definable banks is also without final destination.

Time Hypostatization. Many of the above assumptions are a result of what logicians refer to as time hypostatization. To hypostatize time is reification along with the creation of qualities and attributes. The neo-classicalist reification of time is confusion between time and the experience of succession or sequence.

Time Spatialization. In conjunction with hypostatization, it is also necessary to practice time spatialization in order to see the neoclassical time streams flowing ever outward. To spatialize time is to attribute spatial qualities to time or to confuse the spatial relations among things with the temporal relations among events.

Integrated Definitions. As in Christian time, duration, continuity, and sequence are integrated into one common definition. Time, usually designated by T, is an endless rectilinear stream made up of moments. An event takes place in a moment in T and a sequence of events is spread over the stream of moments. The interval between moments is duration, usually designated by t, thereby defining t as a piece of T. In this definition, reversibility and irreversibility are ignored (which creates real problems for policy concerns like the irreversibility of ecological systems and species extinction).

Space-Time Continuum. Separation of time and space. Although all sciences—from physics to anthropology—recognize the impossibility of separating time and space, the neoclassical discounting formula does not provide for a means to integrate the spatial aspects of the space-time continuum.

Western Time: Modern

To understand the modern concept of time, let us begin by conducting an imaginary experiment. For a moment, imagine that our sensing faculties are suspended in a mental universe of nothingness. Now look out into the empty space. There is no time. Now suspend one stationary item in the nothingness. There is still no temporal dimension. "If we try to imagine ourselves in a world without sound or movement, with nothing stirring, without even our breathing or heart-beats, we must agree that we cannot have Time there. Time may not be merely something happening, but unless something is happening, there cannot be Time."[21] Now let us allow our suspended item to make one move and stop. What was the length of the move? We can see the length of the distance, but there is no measure for the temporal duration because there is no other motion to use as a time clock. If a second moving item is added to this universe, its rate of movement can be used as a measure of the duration of the first item's movement. The second item can serve as a relative time clock. It is apparent that time is relative and duration is relative to another motion. Next, let us fill the universe with items moving in a multitude of directions, in a multitude of spatial patterns, and at a multitude of rates. That fills the universe with potential clocks and different time dimensions. Since these items are moving in a multitude of directions, we might want to say time is multidirectional. However, as stated earlier, that would be incorrect. Time is the duration of the motion. It has no direction. The multitude of moving items is not flowing into the future. In the real world, the earth is rotating around the sun, the bus is traveling about the city, and the barking dog is running around the yard. With this in mind, philosophers sometimes refer to reality as an infinity of layers of now. However, the words "infinity," "layers," and "now" suggest the old ideology. Reality cannot be broken into layers. Layers imply up and down, thus implying direction to time, and we are right back into the original problem of hypostatization and spatialization. In truth, the earth, the dog, and the sun are processing in the here and now. They are not flowing anywhere and neither is time.

Time becomes the duration between the "after" that follows the "before"; for example, one of the dog's feet hits the earth after the other, or one tool development follows the other, and these durations can be measured with some other motion. This certainly takes some of the glamour out of the old concept of future, but, more importantly, it also removes the idea of the inevitability of the future. The old concept informs us that time is passing, and if we do not hurry, time will be wasted and the future will pass us by. As stated above, the modern concept of time allows us to use clocks to regulate events rather than ourselves being regulated by the running clock of passing time. Humans can regulate growth without feeling guilty about cheating posterity. This places human policymakers squarely in the decision-making role of deciding when events should take place. In the old view of time, the relation of "earlier than" and "later than" was absolute and permanent because if an event is ever earlier than another event by a definable interval, it is always earlier, and by that interval. Yet we know that the sequence of events in a system can be changed, reversed, slowed, accelerated, or destroyed. Modern time concepts do not provide for inevitable patterns.

One of E.F. Schumacher's shortcomings, as is true for many other reformers, was his failure to identify the importance of the temporal construct held in the consciousness of the people he was asking to alter their commitment to growth. Because he could not tell us why we felt compelled to do what we were doing, he did not identify the main determinant of the constant push for activity and accomplishment; thus, he could not recommend a viable solution or policy.

Time Analysis Should Be Consistent With Holistic System

In modern thought, time is no longer an exogenous concept, but, rather, another element in the sociotechnical system. When we are planning for a complex sociotechnical system that is constantly evolving, we should not seek a single mechanistic synthesizer of temporal concerns.

Different Clocks for Different Institutional Processes

In order for a social plan to sequence and coordinate social delivery successfully with concomitant "linking-points," planners should be aware that different temporal conditions occur for different kinds of institutional experience. There can be a difference in temporal rhythm and temporal clock from institution to institution. Not only does time change from society to society, it changes among institutions, especially in a complex society. "Time scales may differ greatly over the hierarchical levels of a large system."[22]   One time and one clock do not exist across institutions, as is assumed in neoclassical thought.   One institution may require an even temporal rhythm in the movement of resources, people, and goods. Another may require a relentless pace. Still another may go in impatient jerks, alternating between brief periods with tremendous bursts of intense activity and long durations of plodding activities.

A simple example of different clocks in our society is the example of baseball and the effect of baseball games on transportation delivery systems. Baseball pays no heed to the hour clock, or to Einstein's atomic vibrations for that matter. Baseball has its own special time. It depends on outs, not minutes. Six outs to an inning, rather than sixty minutes to the hour. The duration of a game is determined by the correct number of innings, which may be two to five hours in non-baseball time. This difference in clocks complicates the delivery of transportation for taking the fans home and police for directing traffic while meeting the needs of those on alternative clocks. If the existence of alternative clocks and time is ignored, as in neoclassical cost-benefit analysis, problems are created.

Temporal variation in Brazil is related by Manfred Max-Neef. He explained his frustration in living in a Brazilian town. He was the only person in town who regularly lived by the dictates of the watch on his wrist, while the time of others was regulated by events. "In the short term, by daily events: things are done before or after mass, before or after school classes, after the Town Hall meeting, and so on. The long term is planned and regulated in tune with the religious and patriotic feasts of which there are, of course, a lot. A person's involvement with the preparation of a feast is a duty that takes precedence over any other type of commitment."[23]   Max-Neef has found that the lack of concern for different temporal constructs in Third-World countries can lead to a serious "human state of temporal asynchrony. These asynchronies produce varying degrees of anguish and anxiety, according to the importance given by the person concerned to the bonds of frustrated communication."[24]

However, Max-Neef pointed out that different times need not lead to frustration. To avoid the frustration, it is necessary to be sensitive to, and to integrate the different space-time dimensions in the social system (referred to as "social time integration" later in this chapter). Let us review the integration of alternative space-time that Max-Neef found in the town of Tiradentes, Brazil, because it offers important advice for planners and policymakers. He said:

Time was there, of course, and so was space; but there was something different as well. I had the strong sensation that I was living a "contemporaneousness of the not-contemporary." The mules and the cars, the Chafariz and the television, the sun-dial and my Casio lithium watch. All widely diverging eras co-existing in the midst of a space of incredibly generous perspective. I remembered having been in many old cities before, and my sensation was almost always "time asynchronic": i.e. modern life going on at its usual rapid pace in museum-like surroundings. Here it was different. Times seemed to be synchronized because of the basically tranquil pace and style of the people's forms of human interaction. People were not in a space; they integrated into their space. They defined their own space and made up their own time, thus generating a splendid space-time coherence. It suddenly occurred to me that it was probably very difficult to develop gastric ulcers in a place of this kind. Sometime later I was to discover several forms of space-time disruption, yet this initial impression remained the overriding one for as long as I lived in Tiradentes.[25]

Individual Subjective Time Is Not Appropriate

For economic and social decision making, one of the most dangerous analytical techniques available in dealing with time is the old philosophical technique of turning to individual intuition. "[Henri-Louis] Bergson, an original and bold if rather reckless philosopher, based a whole philosophy on the idea that outer or chronological time is unreal and that reality can be found only in our inner sense of . . . psychological time, the unceasing and creative flow of which we have an immediate apprehension."[26] Nicholas Georgescu-Roegen made this same basic mistake. He said, "there is no other basis for Time than 'the primitive stream of consciousness.'"[27] This approach is laden with problems because intuition is a well filled with our sociocultural groundwater. As Albert Einstein and Leopold Infel said, "The method of reasoning dictated by intuition was wrong and led to false ideas of motion which were held for centuries. Aristotle's great authority throughout Europe was perhaps the chief reason for the long belief in this intuitive idea. . . . The discovery and use of scientific reasoning by Galileo was one of the most important achievements in the history of human thought. . . . This discovery taught us that intuitive conclusions based on immediate observation are not always to be trusted."[28]

The first ability one must acquire is to look outward—not intuitively—for the human being has no internal time or internal time clock. Time clocks are social and time concepts are cultural. Robert E. Ornstein pointed out how early psychologists made the mistake of looking inward for a biological clock (similar to William Stanley Jevons's search for an internal utility calculator). Ornstein stated: "Many psychologists, ignoring that ordinary temporal experience is personally and relativistically constructed, have searched for an internal organ of duration rooted in one biological process or another. This postulated 'organ' has been termed a biological clock. Again, this search follows from the "sensory-process" paradigm of how we experience time, used primarily by those who would try to determine the "accuracy" of our time experience in relation to the ordinary clock. Such thinking confuses, once again, a convenient construction with reality."[29]   Social psychologists find that each year people in the United States have a poorer perception of the rate at which events are happening. That is, they underestimate by a greater amount each year the interval of calendar time between when they expect events to happen and when they actually happen. This psychological phenomenon was an integral part of Alvin Toffler's book, *Future Shock*,[30] in which he explained that futures are coming at us too rapidly for people to prepare for them psychologically.   For the present purpose, such phenomenon certainly should make suspect any attempt to base social planning in a technological society on a personal conception of time or time preferences.

French speleologist Michel Siffre found that after staying in a subterranean cave, which was dark and far out of sight and sound of his fellow humans, without any means of recognizing succession or discovering how clock time was passing, the length of the underground stay was completely misjudged. In Siffre's case, "He found to his astonishment that he had been far longer in his cave than he had imagined. We must note first that no intuitive sense of time had worked for him; that little watchman from the unconscious had gone off duty. . . . Then, his psychological perception of time had lost touch with clock and calendar time."[31]

Some early instrumental philosophers incorrectly based temporal concepts on individual psychological and mental experience. Josiah Royce, for example, dwelt "upon the time consciousness of our relatively direct experience," and concluded that "here lies the basis for every deeper comprehension of the metaphysics both of time and of eternity."[32]   This is inappropriate since complex processes such as those articulated by system networks contain temporal relationships that need to be heeded by the socioecological investigator, whether or not a person experiences them, and the basis for long-term ecosystem cycles is not metaphysical musings about eternity. William James said that the brain process "must be the cause of our perceiving the fact of time at all."[33]   This is true, of course, but misleading if time is attributed to the brain. This is so for at least two reasons. First, there must be an eventful world to perceive before there is time. Second, brain processes can be used to perceive time incorrectly. As anthropologist Edward T. Hall explained, our perception of time "is *not* inherent in man's biological rhythms or his creative drives, nor is it existential in nature."[34]

The notion of psychological time tends to sidetrack any real attempt to solve the problem of social time. First, social life cannot be coordinated with the psychological time of various individuals. The railroad uses clock time for its train schedules—not psychological time. Second, as Priestly stated, psychological time crams into one category too many quite different sorts of experiences.[35]

An additional problem with attempting to develop a succession concept that depends on individual time is that individual time changes with the individual's basal brainwave rate. "All that we experience as external reality is apparently nothing more than patterns of neuronal energy firing off inside our head."[36] Time, therefore, is not even a constant—even given a constant social situation—within an individual brain. In "ordinary" waking moments the brainwave rhythm, beta, is firing at approximately 24 frames per second (to use a movie projector concept). However, if slowed to a firing rate of 10 waves per second, or alpha rate, the second hand on the clock appears to have slowed to approximately half its former speed. If the blasts are slowed to theta, or five blasts per second, time seems to stand still. "When brainwaves are still, time stands still, and when time stands still the illusion of motion becomes impossible . . . to the individual."[37] Time and motion correspond to the individual's current brainwave rhythm. In various states of consciousness, time and motion may be slowed, increased, intermitted, or run backward. By reducing the rate of spheroid blasts, as in meditation, one can fall through the gaps between the firings, and, to use jargon, become "spaced out." Thus, if we are to follow Georgescu-Roegen's suggestion to use stream-of-consciousness, we would first have to know whether it is at the alpha, beta, theta, or delta rate.

Psychological predilections for time preference can be allowed to run free and wild, and even lead and lag, but if social clocks are not coordinated by social time outside and above the individual psyche, then a slight miscalculation of timing, for example, can place one jumbo jet into the path of another at the airport. Individual preference, basal rate, or intuition about time or the speed of the plane is not important. To deliver air transportation safely, pilots and air traffic controllers must be coordinated on a common system.

Timeliness Requires Planning Decisions Be Above the Atomistic Level

In determining the most timely delivery by social programs, it is necessary to take a holistic and organic view of institutions. Depending on individual perceptions, attitudes, and preferences with regard to time and time allocation is very misleading. Individual survey results have no correlation with what institutions actually require. F. Stuart Chapin, Jr., in discussing Stuart Cullen's work on this issue, wrote, "since much of the average weekday is tied up in routines over which people have little day-to-day control, the sequence of a day's activities in the life of an individual is 'pegged' around key structuring episodes. . . . Stuart Cullen sees both practicalities and conceptual problems standing in the way of applying utility-maximizing concepts or in explaining behavior in terms of preference analysis."[38]   Chapin expanded on this issue saying, "rarely do preference studies present choice alternatives in the framework of constraints under which choices must be made. In study after study where preferences are followed up in an investigation of actual behavior, the correspondence between behavior has been of relatively low order."[39]   This result would not surprise holistic social scientists. For that reason, Chapin recommends the holistic approach for sequencing urban public programs. He said, "the view taken here runs counter to the reductionism bias in much of present-day scientific inquiry. But this view is not so much in reaction to these biases as it is a bias in the opposite direction, a strong belief in the necessity of a combinatorial emphasis of a 'whole cloth' view which defines the contingencies of human activity in terms policy makers can recognize, evaluate, and project."[40]

In addition, the individual varies according to numerous particulars such as race, class, recent employment, income, past achievements, and all those matters that color attitudes, estimates, and preferences. However, regardless of class, race, or employment record, the planes need to be on schedule.

From Time Discounting to Timeliness in a Systems Frame

For society to function, it must be an integrated process—integrated into an organized effort by social forces. Before a construct can be designed to assist in temporal evaluation and decision making, it must be understood, as Polanyi emphasized, that society is an integrated system. Policymaking and planning are the processes of instituting new projects into the integrated system so that the right amounts of social goods and services are delivered at the right points in the social system. This calls for Polanyi's concept of sufficiency; that is, earmarking a sufficient amount to fill the needs of the system.

Time, if it is to be a useful concept in social planning, should be the concept of what usually is connoted by the word "timely." Timeliness requires that we ask the question: Which projects will deliver the right amounts of social goods and services at the right points in the social process to allow for the integration, maintenance, and improvement of the social fabric? It is not a matter of the neoclassical "firstest-with-the-mostest" criterion found in present value discounting, but, rather, planning for the coincidence or congruence of the delivery system with need. For example, let us assume that we are able to discern that social forces are such that the future includes a great increase in lung disease in ten years. Timeliness requires that our plan provide for the delivery of the needed care in ten years. That care is not of value today when it is not needed. When planning is done on a substantive basis, rather than a pecuniary one, it becomes apparent that a pneumonectomy cannot be provided today and reinvested for a greater output of pneumonectomies later when needed. The sociotechnical system itself must define the temporal entities, such as time clocks and the temporal sequencing of events. How much, when, and how fast are questions that should be answered by system needs instead of a maximization principle. Therefore, "timeliness" best connotes the temporal concept that should guide project evaluation. If railyards or school buildings are needed at one point in the system evolution, that is when they should be delivered, not at an earlier point in order to increase the dividend found by a geometric mean discounting technique.

Project Evaluation for System Integration

For planning and policymaking purposes, evaluation should assist in making decisions about the coordination of collective social motions and activities, and about the sequencing of events. When the social sequencing approach is emphasized, time is no longer thought of in relation to continuity, but, rather, in relation to duration between sequential events. A time clock is not measuring a continuous stream of moments, but, rather, is a motion that has been chosen as the instrument for measuring the duration of other motions, or of durations between events. To accomplish such temporal evaluation, three concerns must be implemented into the evaluation.

The first deals with one of the main functions of an institution. An important function of an institution concerns the interaction of one institution with another and the mutual reinforcement of the institutions. Thus, the function of alternative policies and programs must be judged to be in accordance with the needs of other system components. The new process should be structured to fit in with all relevant processes, both along and across sequences. How much is needed depends on what is to be provided to other institutions. More is not always better.

Second, and a closely related concern, is the knowledge (as emphasized above in discussing the social fabric matrix) regarding flow levels. "The flows of goods, services, information, and people through the network both structure and maintain . . . community relationships."[41] Formal models often emphasize rates, but real-world systems emphasize the integration of levels. For example, in the economy, the output will be sold if there is an adequate level of expenditures flowing into the market. "Only the values of levels are needed to describe fully the condition of the systems,"[42] because societies organize around levels. "Rates do not control other rates without an intervening level; no rate variable depends directly on any other rate variable. A rate equation is a pure algebraic expression."[43]

Third, consistent with, and necessary for, integrated systems and temporal fit is the structuring of a system without excess surpluses. An excess either of food or industrial pollution is wasteful and often harmful. Surpluses often are the result of flow deliveries that are timed inconsistent with system needs.

Real-Time Control Systems

Temporal evaluation that judges whether a project correctly sequences the delivery of impacts with system needs is consistent with the basic concepts of computer science real time. Real-time systems relate to the sequential events in a system, rather than to clock time.[44] The system itself defines when events should happen. Real-time systems have been used mainly to monitor and control systems by minimizing response time to system deviations. For example, if the monitoring device in a multi-color sheet-fed press (which uses advanced real-time systems) detects a color deviation from the established standard, the control system brings the color back to conformance. Or, as another example, the chemical effusion that prepares plants for winter dormancy and enables them to germinate when the last frost has passed is a real–time system. A failure of this real–time system would cause a plant's extinction. Plants have of necessity evolved appropriate timing mechanisms to ensure their survival. As the length of the day changes, leaves take account of it through the blue pigment phytochrome. Through the interaction of phytochrome and the internal daily rhythms of the plant, there is a change in the production of the inhibitory growth chemical abscisic acid, which plays an important role in the regulatory physiology of the plant. The plant would be in trouble if it attempted to maximize the net present value of abscisic acid instead of coordinating and controlling in a real–time sense with ecological needs. "Control is the process of assuring the conformity of plans and events. Real-time control requires that the response of each element of the control system is such that the combined effect of all elements produces results that are sufficiently expeditious to preclude failure of the system."[45] Harold Sackman emphasized that real-time control is structured around events in an operating system, and does not try to force events to fit into temporally invariant molds.[46] The continual monitoring of real-time control systems "reaches its epitome with computer-based systems, where the computer can act as a built-in laboratory to collect information automatically on how well the system is performing, whether it is achieving its goals or not. But this has to start originally with a plan."[47]

Sackman saw real time as a step toward fulfilling John Dewey's philosophy. He said, "the twentieth century is witnessing the development of a new and practical attitude toward time, an attitude of designing and constructing the fixture through planning and control. Dewey sensed the demise of spectator attitudes toward time and anticipated the trend toward working control."[48] The character of the problem will define the kind of indicator used to measure social impacts. With a real-time approach to temporal evaluation, social indicators, instead of pecuniary prices, can be used as the measurement unit to compare alternative programs.

The real-time concept applied to project evaluation is a step forward in time analysis. However, additional steps are needed to reach the level of social planning and system coordination necessary in the modern technological society.

From Event Synchronization to Social Process Sequencing

With the aid of the general summary contained in Figure 8-1, we can discuss the evolution of time concepts, both in terms of the past and in terms of a needed future. *Time evolution can be summarized as an evolution from event synchronization to social process sequencing.* The evolution of technology is indicated across the top of Figure 8-1; the instruments, or measures, of time are along the left side, and the temporal concepts are arrayed outward from the intersection of the two axes. Figure 8-1 serves as a general summary of the relationship between technological evolution and temporal constructs; as technology changes, both the instruments for measuring time and temporal concepts change. The ordering of time concepts in Figure 8-1 does not mean to imply all the earlier time concepts fade away upon the adoption of more recent concepts.

As stated above, in the simplest technological societies, only a few events had to be synchronized in order to facilitate social life. Time existed only when those events had to be synchronized or when historical occurrences had to be recorded. It did not exist the remainder of the day, week, or year. There were no clocks or a sense of time sequencing. Neither was time divided into units such as weeks or hours. There was no need for such measurements. Sandor Szalai points out that, "in quite a number of poor African and Asian countries the over- whelming majority of the rural population has no access whatsoever to clocks, public or private, and cannot read time."[49]   According to Max-Neef, the same is true in Latin America. He found that "the bond of the peasant with time is different, and has different meanings and consequences, from that of the urban individual, especially one who lives in a metropolitan business industrial environment. There is no doubt that the famous (and very destructive) slogan 'time is money' has no meaning whatsoever for a peasant."[50]

With the evolution of nomadic and agricultural societies, sociotechnical processes became more complex; thus, more synchronization and coordination were needed. Planting, harvesting, and warfare

**Figure 8-1. Evolution of Technology, Time Measurement
Instruments, and Temporal Concepts**

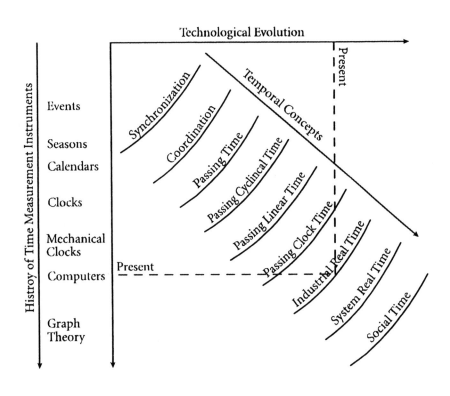

require more refined coordination. In addition, because the seasons
became more important to when and where the tribe moved or when
crops were planted, the seasons replaced events as the main time instrument. This added a natural rhythm and a sense of passing to time as
seasons changed with some regularity. The regularity was noted and
divided into smaller units. Seasons were turned into calendars and hour
dials were developed as the various heights of the sun were recorded.
Even at this, time was not a continuum. "Nobody ever cared in Athens
or Rome to divide the night into hours. What for? With the exception
of a few guardsmen and revelers, the use of time stopped when darkness set in."[51]

The regular rhythm of an organized society gave the sense of
events passing along a time continuum. However, the rate of technological change was so slow that life seemed to be repeating itself from

year to year and from generation to generation. Therefore, the time construct was thought to be circular. As new technological combinations began to appear more rapidly, it became obvious that society was changing. Therefore, the time continuum ceased to be circular and began to move forward, finally becoming linear. The linear continuum became evenly divided into hours and minutes of uniform passing time. "The need for taking account of minutes, not to speak of seconds, arose only when people had already got accustomed to living 'by the clock' and to adapt their daily life, even 'minute' details of it, to the tick-tock of this mechanical device."[52]   With the triumph of linear time, "present time was 'compressed' until it was merely a point sliding continuously along the line which runs from past to future. . . . Present time became fleeting, irreversible and elusive. Man for the first time discovered that time, whose passing he had noted only in relation to events, did not cease even in the absence of events. As a result, an effort had to be made to save time, to use it rationally and to fill it with actions useful for man. The bells from the belfry, ringing out at regular intervals, were a constant reminder of the brevity of life, and a call to great actions which could give time a positive content."[53]

In the industrial era the clock is not just a measure or symbol of passing time. In the minds of that era it is passing time, both operationally and as conscious proof of the passing of time. The clock's 24-hour-per-day, 60-minutes-per-hour, and 60-seconds-per-minute have given the impression of an evenly divided flowing time to those living in an industrial society. Clock time, in conjunction with industrial society, has "established a clock work of its own that imposes its beat even more powerfully on everyday activities than the natural rhythm of the alternating nights and days and of the changing seasons had imposed itself on life in traditional agricultural civilization."[54]   The commitment to and dependence upon the clock by industrial society should not be a surprise because the clock is a machine itself, and, according to Clarence Ayres, "the clock has been called the master pattern of all subsequent expedients in the field of power driven machinery."[55]

Technological advancement allowed for the development of science as well as industry. Since science was emphasizing the relativity of time, the scientific findings were inconsistent with the belief in absolute clock time. We might assume these findings would have made people more conscious of taking control of their own temporal arrangements. However, that has not been the case. "Social life in the industrially developed parts of the world, and especially in modern industrial surroundings, shows an ever-growing dependence on the clock and an ever-growing independence of the calendar."[56]

Although social life continues with clock time, industrial processes are now being changed to real time. This has come about because of the continued expansion of science and technology. It is the result of the integration of holistic science, instrumentalist philosophy, computer science, and computerized industrial production processes. "Contemporary science was criticized as being narrowly compartmentalized around Aristotelian subdivisions of subject matter. The case was developed for science oriented around real-world events and concerned with the regulation and control of events in real time."[57]   As stated earlier, the system determines the measurement instrument. It is not dependent upon the clock of passing time. Real time is defined in a system context that takes account of the appearance, duration, passage, and succession of events as they are interrelated within a system. Real time is concerned with temporary and situationally contingent events amenable to regulation and control. In addition, "it results in an extension of human mastery over such events."[58]   As real time is extended in industrial and computerized systems, its concept will begin to affect more and more spheres of social and personal life. As recommended above, the next step adopts the use of real time for project evaluation. With real time, the computer replaces the traditional clock.

Technological evolution, along with holistic science, will continue to change temporal constructs and extend real time in use and in concept. The extension will be used for socio-technical-environmental space-time planning and coordination. "Relative space is inseparably fused to relative time, the two forming what is called the space-time manifold, or simply *process*."[59]   More and more, it is understood that local, regional, national, and supranational processes must be coordinated and controlled if humans are to solve problems. This will lead to system real time and beyond that to social time, where the events are not just sequenced by the system, but socio-technical-environmental systems are determined by the conscious temporal concept of timeliness through discretionary social institutions.

The Lapps suffered health problems and income losses because of radiation contamination of their reindeer grazing area due to the Chernobyl nuclear plant incident in Russia. The Mexicans have been suffering a loss of water because of the United States' Colorado River dam system, which loses more water through evaporation than it preserves. These are supranational events that could have been foreseen and avoided with the application of scientific inquiry and the development of institutions such as international inspections of nuclear plants and water projects. The inquiry necessary for completing such scenarios was not completed because of the pell-mell rush for development and growth to fulfill the incessant command of passing time. That can change. Time can be tamed, inquiry can be enhanced, and institutions can be created. Human mastery over events needs to be expanded, given the requirements of recent technology.

Space, time, location, and movement cannot be separated. Therefore, these concepts should not be separated in our planning and policymaking. "Without movement, whether tangible or intangible, there are no spatial relations, the main object of study in much of geography. Mere co-existence in space without any effect of one body or process on another is like time without change, or rather like change not producing other changes. . . . Space (and time) seem to be discovered by movement, especially by *locomotion*."[60] When considering events and movement with regard to time and space holistic processes are the relevant context. When it is understood that social processes are policy based, then it is understood that process planning is necessary. For example, there is a global interest, confirmed by treaties and other agreements, in the ducks, geese, and cranes that travel the flyways and staging areas of the United States. Yet the staging areas—wetlands, potholes, rivers, and marshes—are constantly being destroyed, thereby leading to high rates of death and disease among the birds. If we want the birds, then their migration patterns must be coordinated with the environment and with human activity patterns. Events such as drawing irrigation water from the river, providing an adequate water flow in the riparian wetlands, and the arrival of the birds must be sequenced in time, space, and location, or the birds will not survive.

The kind of planning necessary for our transdependent technological system requires a different level than the simple one that just tried to fill the passing time of the clock with production events. We need to adjust and coordinate our beliefs, environment, technology, and so forth. It seems this is a step beyond real time, where the system is used as the clock. This could be called social time, where the system itself is structured.

Traditional time concepts and clocks are not sufficient for the space-time coordination that will take place in the future. "It is not unreasonable to argue that the very fact that we conceptualize human activity as forming a stream flowing through time (and space) suggests that a classification into activity types, sets or modules is either not possible or that it produces a severe and unacceptable abstraction of reality."[61]   In order to successfully plan, coordinate, and monitor, we must change the mode of conceptual notion and abstraction.   The "clocks" of the future are more likely to be akin to computerized matrices, graphs, and digraphs.   The basic reason graph theory is suggested as the likely basis of our time instrument is that graph theory does not conceptualize society as flowing along in a time stream.   In addition, graph theory

> is a general modeling system for relations. . . .   The use of the graph underlying a situation enables us to strip off initially unessential details.   Even when some information is lost in looking only at the graph, this method of modeling can bring new insights by directing one's attention to the structural aspects of the relations being studied.   Where theory exists about the nature of structural relations a graph or digraph theoretic approach provides a means for testing theory.   Where no theory exists or in the more likely situation of ill-defined theories in the social sciences, graphs and digraphs may generate well-defined schemes.   Sometimes it is useful to concentrate on only part of the relation being studied; instead of the whole graph, one looks at a sub-graph formed by deleting some vertices and/or edges from the original.[62]

Digraphs, as explained in Chapters 6 and 7, can be used to represent the sequence of relations and the direction of deliveries among the components of the social system.   Such articulation can be used to plan communication networks, transportation systems, pollution controls, or whatever needs to be coordinated in a timely manner.

Chapters 6 and 7 explained the use of matrices and digraph theory for defining socioeconomic processes and for conducting geobased planning in conjunction with the SFM.   The SFM digraph is consistent with activity sequencing called for by social time.   Activity sequence "puts social events into a system which is ideally timed by the succession of events relevant to that system, that is by *social time*.   In other words, reference to universal or clock time becomes secondary to the internalized timing which is defined by the nature of activity sequence structure.   We have yet to develop a timing system that is internalized to the relationship between events in a socially relevant cycle.   Clock time still dominates."[63]   Sequence methodologies will help us determine and sequence the elements in our institutional structure and process.   "Much of social behavior depends for its orderly qualities on common definitions, assumptions, and actions with regard to the location of events in time. Certain activities, for example, require simultaneous actions by a number of persons, or at least their presence at a particular time. . . . Thus one element of temporal ordering is *synchronization*. Other activities require that actions follow one another in a prescribed order; thus *sequence* is part of temporal order. . . .   For all of these elements of social coordination the term *timing* is useful . . . timing is an intrinsic quality of personal and collective behavior. If activities have no temporal order, they have no order at all."[64]   To order socioecological systems, societies need to leave behind passing time and ultimately move beyond real time to social time. In this way we can achieve the planning and policy necessary for a technological society that is supranational.

Graphical Clocks and Process Synchronicity

Consistent with the potential that graph theory and digraphs hold for temporal analysis, the purpose of this section is to discuss particular dimensions of time, the relations of those dimensions to graphical clocks, and the value of such clocks for network analysis. The clocks for the integration of systems networks will be computerized matrices and concomitant digraphs (graphical clocks). Thus, the network digraph from the SFM can become the system clock and can be used to provide a standardized system clock to determine timeliness. This places events into a common system that is timed by sequenced events. The coordination of broadly recurring sets of meaningful events characterizes synchronicity. Thus, timeliness is defined by system synchronicity. Synchronicity requires special attention be given in policymaking to temporal dimensions.

Process Time

Events in the actual world are process oriented. In terms of the temporality of a process, an event is a happening, to use a term often used by philosophers. It is an ongoing process or part of an ongoing process.

The happening—the verb—creates time by its occurring. The philosopher K.G. Denbigh's conclusion about the noun-verb dilemma is relevant to process time and social planning. "Things which have a material existence present themselves in two somewhat distinct aspects: (1) as being material; (2) as being capable of movement and of change in their qualities. These are reflected in language by the use of noun and verb respectively, and it is perhaps an accident of language that the substantive aspect of things seems to carry with it a greater sense of reality than do their actions or potentialities."[65] Process time is neither reified nor absolute. It is an actual time. "The process does not take any time to occur only because time is created by its occurring. Before its occurring there is no time . . . ."[66] A process approach to time does not attempt to make the world into instants and moments that contain unique events. There is no event that we would call the event in the present moment of time. Instead, there are a multitude of ongoing process flows—flows as diverse as energy and protein flows in wetlands, money flows at retail counters, and weapon deliveries to the war zone. To assume that the present is a unique event in some passing moment, in addition to being an unrealistically narrow focus, ignores process. As the philosopher Henri Bergson wrote, "Real time has no instants."[67] Many philosophers have argued that the belief in past, present, and future is illogical. Without reviewing that literature here, it can be stated that their conclusions follow from the unsubstantiated assumption that time is a flow of moments with present events taking place in a passing moment. In this metaphor, a present event captured in a moment would become a past event as the moment flowed into the future to become the past; this is not logically satisfying.

The recognition that real time has no instants reinforces Stephen D. Parsons' adverse criticism of G.L.S. Shackle's conceptualization of time as a series of instants or moments, "i.e. a sequence of 'nows.'"[68] Parsons argues that the problem with Shackle's understanding of time is that "despite the emphasis on activity, passivity is inherent in the construction of time as a sequence of 'nows' which happen to us."[69] This means according to Parsons, that Shackle was saying, "time passes, it stretches before me as a sequence of 'nows' which I hop along."[70]

The digraph is a tool for conducting inquiry on processes, as explained in Chapter 6 and demonstrated in Figure 8-2. The digraph of

Figure 8-2. Digraph of Overlapping Processes

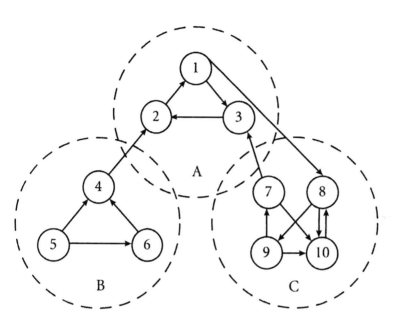

diverse ongoing events provides for unity and meaning, illustrates process sequence, and becomes the process's temporal data. There is not a common mathematics to describe a process because of the dissimilarity of the parts, the transformations, and the complex response and control entities. A process is not a set of simultaneous equations, nor is it a linear transformation. There may be a set of similar ecological entities, for example, a deer population that regularly replicates itself. That replication may be influenced by government policies guided and constrained by human beliefs. Derivatives, integrals, and aggregation mathematics do not allow for the modeling of gaps in common systems such as the socioecological systems that determine deer population. Graph theory and Boolean math does, however, allow for such gaps to be closed in a manner that provides for common networks. We can observe in a digraph (see Figure 8-2) that numerous durations are happening simultaneously. There is not *a* duration (thus, not one time); instead, there are numerous events and numerous durations. Alfred North Whitehead stated that if the investigator chooses one duration as the process clock, a duration is defined as a complex of *partial* events, and the entities that are the components of this complex are thereby said to be simultaneous with this duration.[71]   We shall see later, however, that to select one duration as the process clock would be to ignore the reality of the process approach, that is, to ignore differences in time and time scales among processes.

Process Clocks

Since time is the duration of process happenings, the digraph can be used not only as a description of a space-time process, but also to determine timeliness of the deliveries in the various processes.

Real-time delivery clock. If an event is the delivery of a process flow, the duration is measured by a related flow in the process.   For example, if, during a wheat harvest a trucker is trying to return to the field before the bin of a combine harvester is full, the timeliness of return is determined by the yield of the wheat crop (with a given technology).   In coordinating hauling needs, the flow of grain into the bin is the real flow consideration. The rate of flow of trucks from town to get the grain is a real flow comparison with the grain flow. Trucks must return at a faster rate if the wheat yield is 60 bushels per acre than if it is 30. So the real-time clock is based on trucks per bushel of wheat flow.   As another example, assume that 470,000 gross tons of scrap metal are delivered to a mini-steel mill and that throughout the same duration, the mini-mill uses 780,000 kilowatts of electricity, a real-time clock is 1.7 kilowatts per ton, not minutes per hour.

The use of concepts such as years, hours, and minutes has misled economists about processes because such concepts encourage a perception of flowing time rather than real units. The amount of time occurring during an hour can be increased or decreased by increasing or decreasing the process events and activities. The clock in a digraph is not the edge or line between nodes, as might be suggested by linear time concepts.   Instead, it is the relationships among deliveries that flow among nodes.   Timeliness is the coordination of various flows within a process or among the processes of different systems.

Real systems are polychronic. Edward Hall categorized societies according to whether they had a monochronic or polychronic view of time. He compared and explained the societal consequences of the two different temporal constructs with regard to authority and control, decision making, symbolic images, personal relationships, bureaucratic organization, and so forth.   Hall found that Northern Europeans and

Americans have a monochronic view of life, although they are surrounded by numerous concurrently occurring transactions.   This "monochronic time is arbitrary and imposed, that is, learned.   Because it is so thoroughly learned and so thoroughly integrated into our culture, it is treated as though it were the only natural and logical way of organizing life."[72]   This has led Americans to "think of time as a road or a ribbon stretching into the future, along which one progresses.   The road has segments or compartments which are to be kept discrete ('one thing at a time')."[73]

Paul M. Lane and Carol J. Kaufman applied Hall's concepts to the common analysis of consumption behavior and found that the assumption in neoclassical economics is that people do one thing at a time.[74]   In fact, the neoclassical temporal construct is monochronic to the extreme.   The metaphor of time in neoclassicalism is not only movement along a line (the straight arrow thesis).   Along this line are discrete segments in which events take place one after another, each following its antecedent.   Neoclassicalists carry this reification over into applied time studies by arbitrarily attempting to eliminate overlapping transactions and concurrent activities.   This reification is illustrated by the temporal construct in time-use studies by Juster and Stafford.[75]   Neoclassical time-use surveys force processes into a monochronic, structured clock time made up of neatly segmented, linearly fused blocks of time.   These surveys usually divide a 24-hour day into segments such as 15-minute clocks and require participants to report only one activity in each block in their time diary, regardless of whether the participant is involved in conjoint activities.   The code manual gives specific instructions to eliminate multiple or concurrent activities when respondents are self-reporting.

Unlike North Americans, the Japanese have a polychronic view of time, according to Hall.[76]   Thus, their time surveys report all activities as being done simultaneously.   "In the Japanese data, for example, all activities are recorded even when they are being done simultaneously, with the result that the total amount of time allocated to different activities comes to more than a 24-hour day."[77]   The Japanese respondents recorded according to reality, but Juster and Stafford aggregated as if activities can only take place one after another along a 24-hour line. They warn that to achieve the 24-hour day from the Japanese data "takes a good deal of arbitrary reallocation of time uses . . . ."[78]   Their arbitrary method applied to the Japanese data was to reduce all time uses proportionally except for the total time reported for sleep and for market work.[79]

Monochronic survey design techniques create a bias toward reporting physical activity.[80] This reinforces the materialistic conceptuality of Newtonian time that still dominates the neoclassical paradigm, which is even more disturbing than the arbitrary treatment of data. It means that these studies are ignoring the most important personal and societal activities such as thinking, discussing, valuing, symbolic analysis, application of belief criteria, moral and ethical decision making, and so forth. These non-physical activities are the center of economic activity.

The activities of persons and families are polychronic, and higher order processes such as those that constitute economies, societies, and ecological systems are even more polychronic. Let me clarify by describing a relevant example of a corporate executive from an international investment corporation driving a car home from work in Lincoln, Nebraska. While driving, the executive is listening to international news briefs on a compact disc that he receives daily, and coincidentally, he is talking on his car phone. If a time surveyor requires the executive to list only one activity, he may list driving home and ignore the concurrent processing of news information and the investment deals being made on the phone. If the surveyor requests the executive to list all three activities, and if the drive home takes 30 minutes, then the driver has 90 minutes of activity in a 30-minute period. This is contradictory. This is a contradiction, however, that we need not worry about resolving because it is an artifact of the monochronic approach. Additional complexity can be used to clarify. The three activities mentioned do not indicate all the processes in which the driver is involved, nor do they indicate which processes are of interest. Other processes include: (1) the many physiological processes that are processing and generating durations—coronary, kidney, blood circulation—to mention a few; (2) the auto fuel consumption process with its far-flung accompanying processes of pollution, automobile manufacturing, and Middle Eastern oil production; (3) the family gathering for the evening meal; (4) suburbanization; and so forth. Process intertwined with process intertwined with numerous other processes—that is the real world of driving home.

Even if the monochronic approach to time is abandoned, the question of which time-generating process to study remains. The answer, according to the instrumentalist approach, is determined by the problem we are attempting to solve.   If the problem is one of species survival in an area of Africa populated by tribes of hunters and gatherers, then the activity of the driver in Lincoln, Nebraska that is of interest is the phone call in which the executive concludes a deal to introduce cattle into a new area of Africa.   This will mean the destruction of species that compete with cattle and the consumption of vegetation by the cattle.   Through this process, the sources of food for hunting and gathering tribes are destroyed.   Thus, the events and durations are generating times of doom for many species and tribes.   As John Dewey clarified, time analysis is dependent on the aspect of the problem that is to be considered.[81]

When we turn to study real-world problems, in real-world social processes, the monochronic approach of neoclassicalism serves as a misleading reification that leads, in turn, to debate about other misleading reifications.

> Much of the debate about the nature of time is a fruitless debate, arising from the reification of time. We often treat it as a concrete thing. If we did not, we should not get into arguments about the passage of time, or the flow of time, or about the future flowing into the present into the past, or about whether the arrow of time is unidirectional or bidirectional or directional at all, or whether there is a time arrow, or about the possible effects of entropy upon time. These formulations confuse the concept of time as a positional noun and as a universal category, with the experience of the concretely particular events or processes involving concrete particular things; which are the phenomena from which we construct the concept of time.   It can be stated quite simply that time is neither like a river nor like an arrow. *Tempus fugit* is not true . . . nor does time go in any particular direction, because time per se does not go anywhere or point anywhere.[82]

With real-world problems, reification is not sufficient.   With real processes, numerous nodes are simultaneously transacting different deliveries and thereby creating numerous different time dimensions.   The temporal character of processes is polychronic and multidimensional.

In most cases, economists who attempt to integrate time into the analysis of expectations ignore the polychronic and multidimensional character of time.   An example is an article by Randall Bausor that concerns the expectations governing choice.[83]   Bausor assumes that "time dictates behavior"[84]; yet, as emphasized above, behavior creates time.   At a more basic level, he assumes that time has a "forward motion" with "moments" and "instants"[85] in a world where "causation runs strictly forward"[86]; yet, time does not flow into the future (or into any direction), nor is it constructed of moments and instants.   In the final analysis, however, Bausor's work does not depend on time.   This point can be demonstrated with the following sentence from his article: "It is with the unreasonable reactions from the heart that historical analysis must be most concerned, for they constitute the central events of kaleidics and account for the most amazing distinctions between historical-time and logical-time analysis."[87]   The word "time" could be left out of the sentence without changing his intended meaning.   Indeed, to do so would clarify the meaning.   If we, however, actually introduce temporality by substituting the correct synonym for time, the sentence loses its meaning.   If "duration" is substituted for "time," the sentence is confusing.   What is "logical duration"?

The Multidimensional Temporal Character of Processes

For polychronic synchronization to provide order and timeliness in a process, temporal differences within and across processes need to be recognized.   Temporal differences that complicate the maintenance of synchronicity will be discussed next.

   Real time; dynamic change for stability.   Events and deliveries must change on a real-time basis to maintain a system.   This is part of a more general temporal principle often ignored in economics, a discipline in which a duality is often maintained between dynamics and stability.   In real processes, dynamic change is necessary to maintain stability.   In automobiles, the carburetor governor changes the flow of fuel as the load on the engine changes in order to maintain a stable speed.   In the modern economy, changes are often made to maintain a stable economy.   These include legislative changes in tax policy, central bank changes in the cost and availability of liquidity, and the operation of automatic stabilizers that respond to change by changing flows such as unemployment compensation payments.

   Real-world processes are not equilibrium systems.   Thus, decisions, responses, and reflections are made on a real-time basis to maintain stability and the re-creation of processes.   Dynamic control must change particular events, reactions, and trajectories relative to other un-constrained changes. The collection of elements and components (represented by the nodes in a digraph) persists largely because of changes in their deliveries in response to other deliveries.

Different processes.   Different processes operate at different temporal rates. Crucial to coordination and order is the regulation of delivery from an adjoining process in a manner consistent with the temporal frequencies of the receiving process. This concept is represented in Figure 8-3. If the system processes A and B in Figure 8-3 are

Figure 8-3. Process Digraph of Different Process Frequencies. High Frequency of B is Modified for Low Frequency A

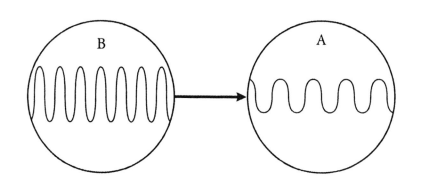

viewed as whole processes, without respect to their internal pattern, B is represented to have a higher frequency within its process than A. Thus, the frequency of B cannot be left unmodified in its deliveries to A because, otherwise, A would be disrupted. Figure 8-3 illustrates that a low-frequency process needs to be isolated from the high frequency of another process. Several examples come to mind. One is the United States Supreme Court's decision with regard to market transactions in the economy. Market transactions occur at a very rapid pace, while the court issues opinions regarding market transactions at a much slower rate. The court would be overwhelmed if it did not extricate itself from the high-frequency signals of economic transactions. Another example is the relationship between ecological systems and economic systems. Ecological systems operate on large, slow-moving time scales. Their response time to impacts and disruptions is slow. This means that without managerial control to protect ecological systems from the rapid pace of economic systems, the former will be overwhelmed and destroyed. Thus, "environmental management should focus on trends in the behavior of parts and that 'ecological breakdowns' such as the dust bowl occur when rapid changes in social trends overbound the parameters of normally slower change in larger systems."[88]

Successional time. Succession adds another time difference. Economics has long been concerned with the succession of alternative conditions and appearances. The business cycle, with its succession from recession to recovery to boom to recession, is one case. The financial stages of a business provide another example. Although the technology and production process may remain the same, a company may pass through numerous financial stages, beginning with seed capital, through periods of negative cash flow, to positive cash flow, and finally to the need for expansion capital.

Although common in economics, consideration of the temporal dynamics of succession did not take place in ecological studies until about 1900. The succession from weeds to grass to trees following ecological disruption is an example familiar to most of us. In Midwestern prairie pothole marshes, the vegetation can be wiped out by a population explosion of herbivorous muskrats. New emergent vegetation cannot become established until a dry year exposes the soil. A typical succession of species follows until robust cattails out-compete them. This sets the stage for another muskrat explosion. This cycle has a six- to eight-year frequency that depends on both the biotic vegetation-muskrat interaction and the biotic wet-dry climatic cycle.[89] Successional time has played a central role in ecological studies. However, like the economy, the ecosystem cannot be explained fully by successional time.[90] There is also evolutionary time.

Evolutionary time. The time dimensions of a process are based on the sequences of events and on the succession of dynamic processes. Evolutionary change, however, connotes a magnitude of change such as to alter time scale. An evolutionary clock is neither locked into a sequential time frame of particular components and deliveries nor into a succession of alternative collectives. *All* is in flux, including time. Evolutionary change can also be expressed through digraphs. Figure 8-4 is presented as an evolution of Figure 8-2.

Figure 8-4. Evolution of Figure 8-2 with Acquisition and Loss of Nodes and Deliveries. Loss of Nodes Indicated by Shaded Circles. New Deliveries Indicated by Dotted Lines. Nodes Lost are 3, 9, and 10. Nodes Gained are 11, 12, 13, 14, and 15.

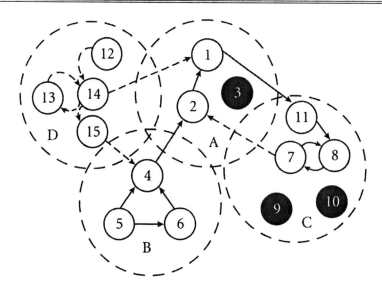

Figure 8-4 illustrates that the system (1) has become more complex, (2) has evolved new nodes and deliveries, and (3) has lost particular nodes and deliveries. This evolution creates new times and new durational relationships. A well-monitored digraph allows us to track and evaluate evolutionary change. The complexity of a system digraph depends on the extent of the evolution under consideration. The greater the evolutionary activity, the more some components die out and new components come into being in the system; the less the activity, the less complex the digraph. Concepts such as evolutionary economics, ecodevelopment, successive eras, development, historical stages, cumulative circular causation, coevolutionary development, and growth can be expressed and applied much more accurately through comparative digraphs. In addition, such expression allows for the conversion of the digraph to its Boolean matrix in order to take advantage of additional analytical techniques.

Social Time

Thus far, temporal coordination has been discussed in terms of "real time"; that is, the coordination of active real-world processes. Instrumental policy, however, is concerned with changing real-world processes. In an advanced technological society, humans become the gods. Knowledge of past policy errors may make intelligent analysts want to reject the policymaking task of synthesizing socioecological change. Reality, though, forces us to recognize that for better or worse, assigning tasks to create such change is a human responsibility. Policies determine which species are to exist, which children are to be hungry, how soon the ozone cover will be depleted, how high the incidence of cancer will be, how high the flow of income and investment will be, and so forth. Thus, the task goes beyond simply coordinating processes; the task is to determine them. With this recognition, a step is taken beyond real time to social time, through which societal decision making determines order, timeliness, nodes, and deliveries. In terms of SFM and digraph articulation, the task is one of design *and evaluation* of alternative system matrices and digraphs.

Conclusion: Instrumental Policy and Time Analysis

Adequate instrumental analysis cannot be completed without accounting for the temporal structure of processes. John Dewey stated that "the nature of time and change has now become in its own right a philosophical problem of the first importance."[91] Yet, a perusal of policy science literature provides few examples of use of temporal analysis or of how temporal concerns have been handled in particular studies. Temporal analysis needs to become more explicit and form an integral part of policy studies.

Examples of Carelessness

Some studies, many of which are not explicitly allied with the Newtonian materialistic concept of time, are replete with terms and phrases that associate closely with that concept. One example that is common is the treatment of "historical time." This concept is based on the assumption that time is passing, although time does not in fact pass.

Treating time as space also appears. An illustration is the term "over time." Is it possible to be over that which is not spatial? Time is not spatial. Time is found because of process; process is not found "over time," or "in time," or "through time." To make time prior to process is to make time a metaphysical enigma.

Some phrases disassociate time from events and duration. One such phrase is "a point in time." Yet real time has no points. There can be time only if there is duration between events and there is no duration at any point. A similar case can be made against the phrase "at a particular moment in time." Another that makes the same disassociation is the phrase, "Time is money." Time is not money. Monetary processes create time; time does not create money. A similar disassociation is made with the statement, "fiscal policy takes time." Fiscal policy does not take time. It creates time. As Whitehead stated, "The disassociation of time from events discloses to our immediate inspection that the attempt to set up time as an independent terminus for knowledge is like the effort to find substance in a shadow. There is time because there are happenings, and apart from happenings there is nothing."[92]

Sequences of Events, Networks, and Timeliness

Matrices and concomitant digraphs can be utilized to complete the description of social networks, as has been the case in sociology and anthropology. As stated above, such digraphs can be utilized as graphical clocks. The value of such graphical clocks for policy research is in order to determine timeliness in the description and evaluation of socioeconomic systems. In sociology, where most of the network analysis has been completed, the networks have not allowed for succession or evolutionary temporal frames. The digraphic techniques suggested above will allow scientists to complete the successional and evolutionary analyses.

Much of the research important for policy analysis is dependent on the concept of timeliness. A description without an understanding of the timeliness of deliveries to the nodes in a real process should not be considered complete. Likewise, analysis of either (1) the agents within a system or (2) recommended policy requires an understanding of what, when, how much, and how far. Discretionary decision making requires providing for the real-time dynamic response to provide for process stability, as described earlier.

Technological Change

Technological change is an important factor in real-world systems. Its importance to time and timeliness needs to be recognized and incorporated into policy analysis. As noted in the wheat cutting example above, real-time delivery clocks are dependent on a given technological context. A change in technology alters time (alters the events, durations, and happenings). With the change in activities, the meaning of temporal concepts such as person hours and person years changes. An example of change in temporal characteristics due to new technology is provided by global wireless communication systems. The events happening changed, the frequency and rate of such events changed, the duration between events such as the exchange of money was decreased, and processes such as payment and monetary processes changed. Electronic money so changed the temporal rate and ease with which money can be transferred around the world that the formation of new global regulatory institutions was necessary for the continued relevance of policy. These kinds of evolutionary changes can be articulated by the use of graphical clocks as was illustrated in Figure 8-4.

The Social Fabric Matrix for Temporal Processes

Time is based on duration, and clocks are events during a duration used to measure and coordinate other events. There are numerous times and real clocks in a process, as there are numerous ways that a process might be ordered and coordinated. Digraphs and their matrices can be used as graphical clocks for understanding and synchronizing the order and timing of processes. The common time unit is the network (digraph) of the system. *Within a process*, events are the deliveries among process nodes (institutional, technological, cultural, and ecological). *Process stability* is maintained through dynamic process change. *Process succession* is the succession of alternative conditions and new collective sets. Business cycle conditions and ecological succession of species are examples. *Process-to-process* deliveries require frequency adjustments to make deliveries consistent with receiving processes. *Evolutionary change* means a drastic change in the process components and its deliveries, dynamics, and succession. All of these process occurrences create a complexity of temporal dimensions. To articulate the complexity is crucial in order to describe, understand, or plan for a process, and graphical clocks derived from SFM digraphs are useful instruments for such describing, understanding, or planning. The SFM and digraphs provide for common networks that will permit policy analysts to computerize the data bases that describe processes, to model whole socioecological collectives, to determine the thresholds of deliveries, to monitor change, to allow for real-time order and response, and to conduct network evaluation.

CHAPTER 9

EVALUATION FOR SUFFICIENCY: COMBINING THE SOCIAL FABRIC MATRIX AND INSTRUMENTALISM

This chapter draws on the ideas presented thus far in order to explain the use of the SFM approach to complete an instrumental evaluation that emphasizes sufficiency. Ecological examples and concerns are used as an explanatory vehicle. Ecological studies have emphasized a concern for sufficient flows. In addition, policy ecologists have recently clarified that particular policy principles such as the precautionary and sustainability principles require a complex systems approach that encompasses social, technological, and ecological components. Traditionally that has not been the case in ecology. Although ecological policymaking concerns are emphasized in this chapter, the main purpose is to present a sufficiency approach to evaluation.

Discussions regarding ecological crises often begin by casually identifying human actions responsible for such crises and by briefly explaining the need for holistic or macroscopic modeling. Yet, when modeling begins, policy scientists, philosophers, and social scientists usually are not included, and the models often are not ecological, but are rather narrowly biological or physical. This chapter will emphasize the need for broader and richer modeling and the need to recognize that environmental protection and enhancement is driven by human policymaking in a sociotechnical context. The question is not whether it will be human policymaking, but rather which values, beliefs, and philosophies will guide the policymaking paradigms and analytical techniques.

The Need for Context and Criteria

Deliveries and consequences between natural ecosystems and social organizations are the means for each other so that neither can be under-

stood nor valued separately from the other.   To think about the qualities of a thing "is to look at a thing in its *relations* with other things,
and such judgment often modifies *radically* the original attitude of es-
teem and liking."[1]   An entity that is valued highly is one "which serves
certain ends, that which stand in certain connections with consequences.   Judgment of value is the name of the act which searches for
and takes into consideration these connections."[2]   Economic and ecological professionals have often ignored Dewey's advice regarding deliveries, connections, and consequences among related entities, so that
most standards and principles of ecological valuation are based on what
Alfred Whitehead called the fallacy of misplaced concreteness.

For policy evaluation to be useful, policy scientists must be
context specific.   A contextual approach means that an investigator
does "not flatten out all decisions into interchangeable units of individual welfare, but instead retains a sensitivity toward different types and
scales of impacts."[3]   The flattening-out approach to evaluation—the
measuring of welfare according to a single scale of indicators such as
dollars—ignores differences in the different kinds of consequences."[4]

In addition to criteria and context being needed to model and
analyze, Herman Daly and John Cobb suggest two rules of advice from
Whitehead about how to avoid toxic levels of abstractness.   The first
is, in Whitehead's words, "recurrence to the concrete in search of inspi-
ration"[5] and the second is to avoid excessive professional specialization.

Concern for criteria, context, and concreteness becomes a call
for a change from the traditional maximization metaphor.   The recommendation here is to approach policy by regularly inquiring about
what is appropriate, or, as Aristotle advised, the purpose is the achieve-
ment of the good.   The concern for concrete criteria applied in a concrete context requires a different metaphor.   It is a metaphor that includes limits, sufficiency, policy relevance, and multidimensionality.
Changing the ideological metaphor allows recognition that welfare is
increased by decreasing options, by limiting production, and by directing technology.   This allows recognition that investment spending and
increases in GDP can reduce welfare.   It is a metaphor that emphasizes
improvement rather than growth.   To improve systems requires that
particular attention be devoted to the sufficiency of flows to provide for
a good system.

Integration of Social and Ecological Paradigms

In her book, *Crafting Institutions,* Elinor Ostrom explains that many economic infrastructure projects around the world have failed because institutions were ignored.[6] Both the analysis of current institutions and the crafting of new institutions are necessary to make projects function. Project modeling has usually been the exclusive domain of engineers and cost-benefit analysts, to the detriment of projects. Although Ostrom's main concern is with irrigation projects in Third World countries, the same kind of problem exists with regard to the modeling and analysis for ecological policymaking. Most ecological analysis is completed by ecologists and neoclassical economists. The two usually ignore each other's analysis and they both ignore institutions and the design and crafting of new institutions.

Let us define an approach for a relevant decision domain in which social institutions and ecological systems both operate. Such context was broadly illustrated above in Figure 2-2 and explained in Chapter 6. The components in Figure 2-2 were explained as being relevant to all problem areas. The importance of the Figure 2-2 digraph for the purpose here is twofold. The first is to demonstrate that a purely ecological problem does not exist. It is embedded in larger sociotechnical systems. A second point revealed in Figure 2-2 is that technology delivered and applied to the ecological system is delivered through institutions, and the substances and services provided from the ecological system are delivered to institutions. In one sense, this simplifies modeling by clarifying that direct deliveries do not occur between ecological systems and other sociotechnical components except through institutions. In another sense, it complicates analysis because searching out the relationships among other components that indirectly influence the institutional-ecological nexus is often difficult.

Additional complexity obtrudes itself upon the analysis because of our knowledge about institutions and the ecological system. There is never a direct relationship between any one institution and one ecological component. No institution operates alone. Institutions overlap with institutions in any social endeavor; thus institutions work through institutions in transaction with ecological components.

The wetland model articulated in Figure 9-1 (and it is the predominant kind in the literature) is incomplete.[7] No one can find a wetland that is separate from human institutions (or articulated differently,

Figure 9-1. Wetland Ecosystem Model

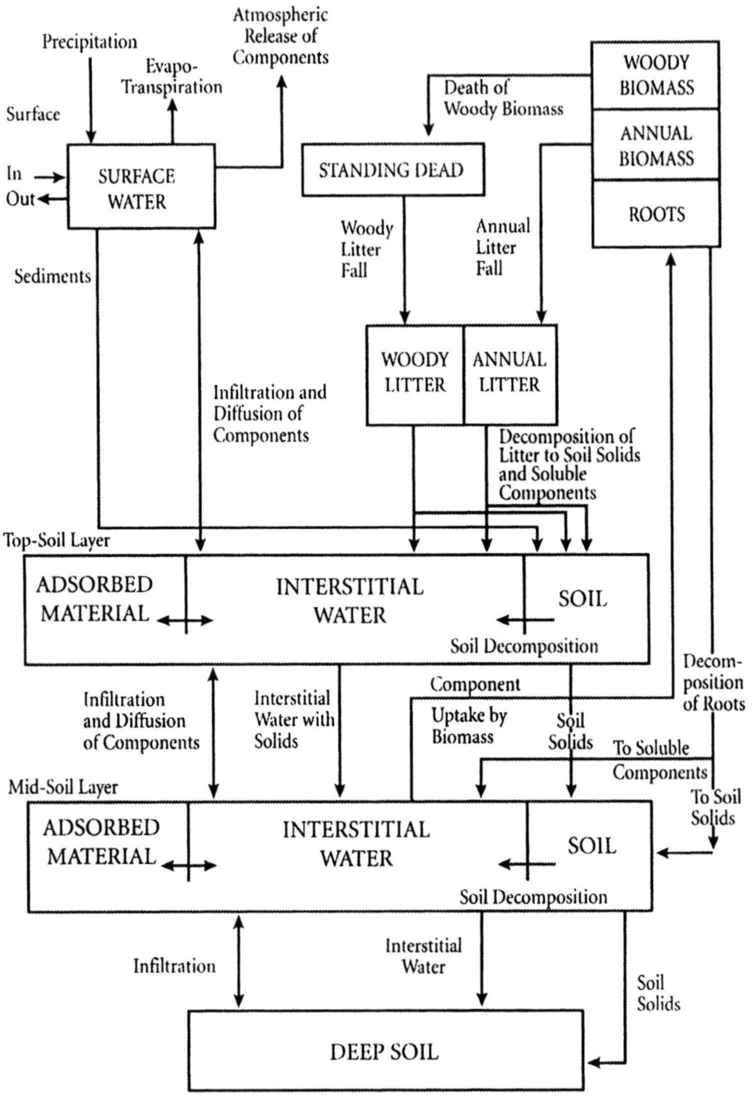

Source: Kadlec and Hammer 1988.

and more correctly, real-world wetland systems are not isolated natural systems). It is not possible to isolate a wetland as a system and then take account of the impact of an "outside force" on it as a system. First, "the force" is really numerous and varied forces, or deliveries.

Second, they are not outside because institutions do not operate apart from ecological systems. No institution ever existed apart from ecological systems. Most deliveries from ecological systems are taken by the institutional application of technology. Thus, *one of the main tasks of ecological policy is crafting appropriate institutions*. Third, each of numerous institutional entities is likely to be related, in terms of delivery, to a number of different kinds of enecological entities. An example is the stress from a xenobiotic organic chemical delivery that is toxic to biota and associated with 19 different problem areas. The association includes *various institutions* and their deliveries, such as mining wastes and agricultural pesticides; *numerous mediums* for delivery, such as sludge and ground water; and *diverse receiving nodes*, such as wildlife food sources and human water sources.

If we recognize the regular exchange between the ecosystem and social institutions, then concepts such as ecological sustainability and biodiversity become much more difficult to define separate from a particular context and much more difficult to apply in a context. Ecological sustainability has a social meaning because ecosystems have a social context, as is also the case for biodiversity. Alternative ecological sets need to be defined and considered for each alternative context. It is not sufficient to think that the same wetland will be treated differently depending on different technological and social habits. The wetland itself changes with each different technological and social regime. The kinds of ecological functions and species to be sustained vary as the socioecological process changes. God is not the only concept that is in the details—to paraphrase Einstein—so are wetlands. If gold mining tailings are dumped in a wetland, the details have changed. The kind of chemicals and their form can change with each different transactional context. To answer questions about biodiversity and sustainability, human groups, their technologies, and their institutions are significant and need to be so recognized in modeling.

When modeling, with a concern for sustainability, contextual specificity requires knowledge about which cycle is relevant. What point in the cycle is being modeled? Will an attempt be made to freeze the ecological component in its current cycle? In a six- to eight-year cycle of biotic vegetarian-muskrat alteration, muskrat sustainability efforts can only coincide at one cycle in the alternating succession. Under conditions of autogenic succession, wetlands in a natural setting are completely destroyed. Wetlands gradually fill because of erosion of mineral material and the accumulation of organic material. Eventually, the water becomes shallow enough to support marsh vegetation and build a peat mat.[8] Thus, an attempt to sustain a wetland and all its immerse array of life is "unnatural." (It may be that the only conditions under which constant sustainability can exist are those under anthropocentric control such as the United States corn belt, where, at considerable expense, a particular kind of ecosystem is sustained.)

The edges (deliveries and connections) in a SFM digraph, which are to be defined for policy purposes, should be defined in order that sufficiency flow decisions can be determined. Figure 9-2 can be used to visualize the meaning of a sufficiency delivery concept. Assume factory J in Figure 9-2 is delivering pollution to wetland K. In a

Figure 9-2. Delivery Criteria Indicies (m and n)

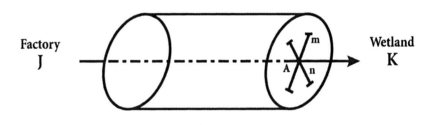

sufficiency model, the delivery carrying capacity would be limited by what is sufficient to meet the contextual criteria of the process. For simplicity, assume that the dimensions of concern are m and n. If m and n are consequent indexes, they should be designed to indicate the results of the factory pollution on a species ecosystem function or ecosystem service. Although m and n are different dimensions of the flow from J to K, m and n can be normalized around a common zero point at point A. The common zero point is the flow level that has been normalized to make a particular system viable. Determining what level of chemical discharge into a wetland is tolerable depends on the system to be sustained. The zero point indicates zero deviation from the ecosystem normality that decision makers are attempting to achieve.

Let us assume the spectrum index across the top of Figure 9-3 represents m (from Figure 9-2) and the similar index on the left side of Figure 9-3 represents dimension n. They have several relevant points.

Figure 9-3. Decision Space With Two Criteria Indices

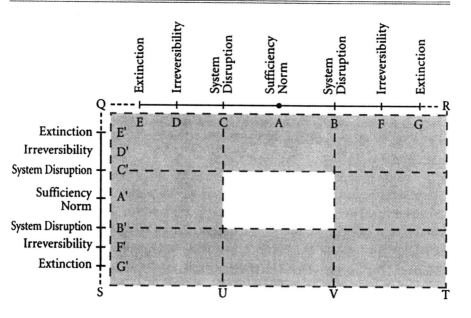

Points A and A' represent the sufficient flow level of a species or eco-system function or service; the flow level necessary for process suffi-ciency. Any point to the left of the norm on index m and above the norm on index n indicates flow levels that are greater than necessary, and any point to the right of the norm on m and below the norm on n indicates flow levels that are less than necessary for normalized suffi-ciency. As in any process, some deviation from the norm is allowed; deviation limits are established to indicate the deterioration of the so-cioecological system. Points B, B', C, and C' represent system disrup-tion without irreversibility or extinction being reached. Points beyond C and C' represent points where too much is being delivered to keep the process viable or sustainable. Points beyond B and B' represent insuf-ficient levels. As an example, the water cleansing and aeration func-tion of a tortuous- turbulent river is depleted when the river is dredged and the channel straightened.

Points D and D' represent the point of irreversibility of an eco-system function or species because of too much. Point E and E' repre-sent complete extinction of an ecosystem species, function, or service.

As an example, excessive numbers of cattle have been introduced in ecosystems around the world, causing the extinction of plant species. Points F and F' indicate a situation in which the socioecological system is providing a flow delivery level that is too small to maintain the environmental function or species. For example, fecal matter flowing from a lake may be blocked by an economic project. As the flow level is decreased, a point is reached that creates severe disease problems for native species. Points G and G' indicate that a species or environmental service has reached the level of extinction because of an insufficient flow.

If only criterion m were applied to QRTS, decisions that lead to delivery flows outside of CBVU should not be allowed. If both criteria m and n, as represented in each index, are applied to the policy decision context QRTS, decisions outside of the white center rectangle are excluded from consideration in order to maintain system viability and integrity. Although drastically simplified, an important policymaking rule is demonstrated with Figure 9-3. In a real-world setting, welfare is improved by limiting decision options and production alternatives. Particular kinds of technology, production, developments, and policies are excluded in order to maintain sufficient flows consistent with the kind of system wanted. This rule is inconsistent with the growth maximization rule inherited from classical—whether capitalist, Marxist, or socialist—thought, which was a one-dimensional growth ideology that judged welfare to be improved by increasing decision options and production alternatives.

A complementary policymaking rule is that planning will be improved if policy delivery criteria are made explicit for decision makers. Again and again corporations made technology decisions without explicit norms and limiting sufficiency spectrums. After the production process is constructed, it is determined that the flow of toxic materials is too harmful for the human genetic system, for example. Thus, the technology needs to be changed, and the production process retrofitted. This is an extremely wasteful approach. If the limiting criteria are made explicit before the innovation and planning process begins, the planners and entrepreneurs can be creative within the limited decision space allowed. In this way, technological innovation will be consistent with system needs rather than destructive of the system.

Ideological Impediments

George Lodge, and others, have taught us that it is important to explain the ideological impediments that prevent appropriate analysis. Progress in designing transactional ecosystem models has been so meager because paradigms in both economics and ecology have been guided by the belief that some universal presence leads to sustainable equilibrium through maximization.   In an article about ecological communities, Richard Levins wrote that "despite assertion that communities evolve to maximize stability or efficiency or information or complexity or anything else, there is no necessary relation between evolution within the component species and evolution of macroscopic community properties."[9]   Yet these claims continue to be made and are attractive to economists and biologists.   "Perhaps the reason for this is a frequent reference by biologists to a philosophical framework that seeks harmony in nature.   Or it may be the transfer to ecology of the equally invalid Adam-Smithian assertation in the economics of capitalism, that some hidden hand converts the profit-maximizing activities of individual companies into some social good."[10]

Adam Smith was expressing a more general Western myth. The Western mind holds a whole set of noninterference beliefs, examples of which include: (1) "Natural species provide for a sustainable environment"; (2) "Do not interfere with the market lest chaos be created through disequilibrium"; and (3) "Mutation will provide the variation for natural selection to favor."   The noninterference equilibrium model of the invisible hand is so pervasive that it is sometimes used to create a common denominator between animal behavior and economic institutions.   This was demonstrated in a federal research grant application.   The university professors who wrote the application asserted that work on foraging behavior in animals in the laboratory has provided good test models for complex theories of economic decision making.

With the ideology of sustainable equilibrium, systems are metaphysical enigmas that are guided by invisible supernatural forces.   The mystery of life is solved.   The complexity and dynamics of open living systems are replaced by elegant simplicity.   Efficiency in an environmental system, as in the market system, is achieved by the same policies.   In ecology, as with Adam Smith, the more competition, and the more diverse the competition, the better for the system.   More species and more diverse kinds of species; more firms and more diverse kinds of production—both configurations provide for efficiency.   Instrumentalists have taken issue with the assumption of the invisible hand and sustainability theories in economics.   Equal suspicion should be cast on the same kinds of theories in the biological sciences.

Levins reports on studies in dynamic biogeography that have not found sustainable equilibrium in natural settings.   Biologists have found surprisingly high turnover rates of species on islands and other isolated patches.   They also have determined that "the turnover rates of species depend on their whole environment, including the structures of the communities in which they live and on the genetic make-up of the local populations."[11]

Conclusion

Many researchers in biology and ecology are arriving at the same methodological position as systems analysts;   the study of individual parts will not allow us to understand the parts or the whole.   Neither the parts nor the whole are given by any iron law of nature in either ecology or economics.   They are more and more given by technology and institutions.   Humans must decide upon the kind of technology, institutions, and environment.   Therefore, we are led back to instrumental philosophy to decide what to study and how to study it.   For instrumental valuing, concerns about criteria, concreteness, and context come to the fore.   They need to be resolved into system sufficiency norms, and sustainability and biodiversity need to be instrumentally judged in the context of the whole socioecological process.

Digraphs presented in earlier chapters were used to demonstrate deliveries among components within the SFM.   They can also be utilized to understand that flows among components need to be sufficient if a system is to be sustainable.   Or, if policy is to change a system to a new configuration, the criteria for the various deliveries among the components must be determined in conjunction with each other.   For every delivery edge in a digraph, there will be one or more sufficiency spectrums as articulated in Figure 9-3.   Through sufficiency criteria, policy can be made for sustainable systems.   A task of instrumental policy analysis is to evaluate for sufficiency norms and design relevant

indicators to represent those norms, as well as to measure deviations
from the norms.

CHAPTER 10

THE SOCIAL FABRIC MATRIX IN A METAPOLICYMAKING CONTEXT

The SFM approach to inquiry must function in a context of policymaking that includes politics, lobbying, scientific analysis, the budgetary process, and so forth. It must fit into the policymaking process to be relevant, and, furthermore, it must function in the public arena in order for analysis to be improved. "The tools of social inquiry will be clumsy as long as they are forged in places and under conditions remote from contemporary events."[1] Consequently, this final chapter has two purposes. The first is the explanation of a transdisciplinary and integrated approach to metapolicymaking. Metapolicymaking is concerned with policymaking regarding policymaking. The second is to place the SFM approach in the context of metapolicymaking.

Figure 10-1 is offered to provide an overall view of policymaking. It is not intended as a model, a linear sequence, or even a complete taxonomy. It is offered to outline the most relevant phases and levels of policymaking. Across the top of Figure 10-1 are the phases of policymaking, and from the top down are the levels. The levels are policy, strategy, and tactics, as indicated on the left-hand side of Figure 10-1, with their respective sciences indicated on the right-hand side. The lines connecting the boxes are there to indicate that for policymaking to be effective, all phases and levels must be consistent and integrated, not to indicate a mechanistic lock-step operation. Policy scientists need to fill the 30 boxes contained in Figure 10-1 with tools and integrate them in a complete policymaking process. No one scholar or policymaker can be an expert in all the areas; each box is an area of study and expertise.

The more northwesterly area of Figure 10-1 is where most academics work; toward the southeast corner, policy is finalized and implemented. For effective planning and policymaking, all corners need to be mastered and tied together through the integration of all areas in between the corners. If a policy scientist laboriously cultivates the

Figure 10-1. Policy, Strategy, and Tactics of Policymaking:
Phases and Levels of Metapolicymaking

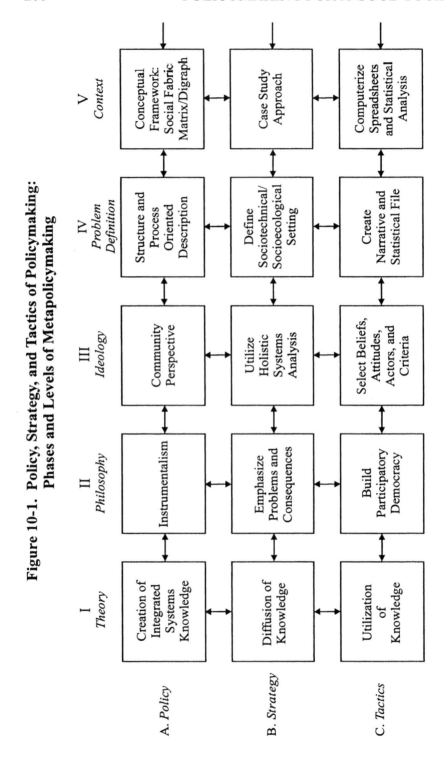

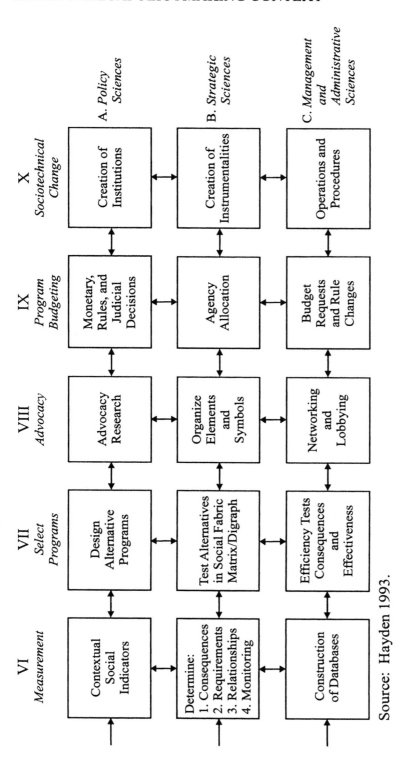

Source: Hayden 1993.

knowledge vineyard in the northwest quarter without the means to integrate the knowledge created, the work will not bear fruit in the policy field. In order to avoid such frustration, it is necessary to close the gap between science and tactics–between theory and operations. In addition to the production of science and philosophy, there is need to produce strategies and tactics. That means politicians, lawyers, planners, lobbyists, economists, accountants, budget analysts, and the like with a systemic and transdisciplinary mindset must also be recruited.

The purpose of Figures 10-2, 10-3, and 10-4 is to indicate three serious and prominent policymaking problems that occur when the message of Figure 10-1 is not heeded. Figure 10-2 is labeled the "Bureaucratic Approach" to policymaking. In this approach, tactical and administrative personnel make policy or conduct their activities without respect to the knowledge base, philosophy, problem research, belief system, or policy strategies. This approach is usually guided by techniques acquired by experts without inspecting the techniques to determine their appropriateness. "If the expert is to be useful at all he must be integrated into a general scheme and led by a generalist who is sensitive to the interplay of all the parts."[2]

The second approach, demonstrated in Figure 10-3, is the "Pseudostrategic Approach" in which political strategists: 1) lack the technical expertise of the accountants, psychologists, computer scientists, fiscal analysts, and so forth; 2) do not have the knowledge base of the policy scientists; and 3) are seldom trained in strategic sciences. They usually consider themselves too politically experienced to be concerned with the findings of scientists or the details of tactical expertise, thus generating great frustration for those in a policymaking process who are attempting instrumental policymaking.

The "Scholarly-King Approach," indicated in Figure 10-4, leads to wasted efforts and resources rather than bad policy because the advice from such an approach is seldom heeded. University scholars who devote themselves to general theoretical and philosophical research sometimes direct abstract solutions (often with no more than a page or two of explanation) to government operation and implementation personnel although the "solutions" are not grounded in case studies of the problem area and have not been legitimized and approved though the advocacy process. This kind of suggestion cannot be utilized; thus, it usually engenders no more than a polite reply that often frustrates the scholar.

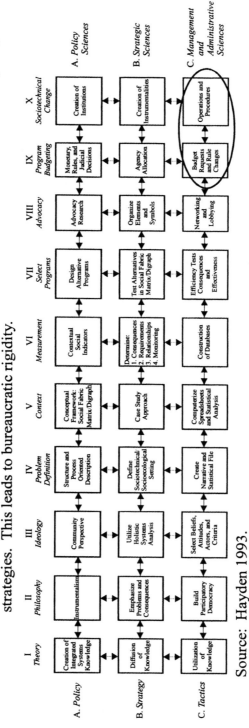

Figure 10-2. Bureaucratic Approach: A "practical" approach concerned only with operations in the implementation phases. Tactics and operations are directed in the implementation phases without concern for higher level beliefs and basic research and without concern for policy strategies. This leads to bureaucratic rigidity.

Source: Hayden 1993.

Figure 10-3. Pseudostrategic Approach: Concerned with designing strategies without being concerned with whether strategies have either a research base or operational feasibility. It is practiced by "political" types who are too "experienced" to be concerned with ideology or research. They listen to the latest charlatan, follow fads, and regularly promote new strategies. If pseudostrategists get a government appointment, they create great bureaucratic scar tissue which must be overcome by serious strategists.

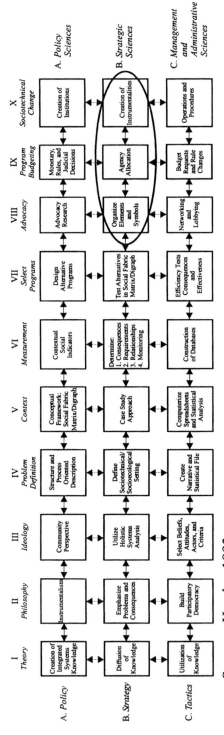

Source: Hayden 1993.

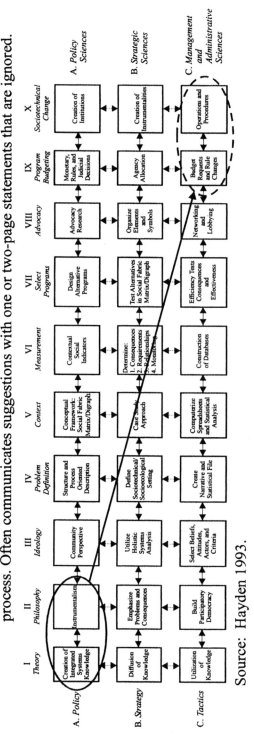

Figure 10-4. Scholarly-King Approach: Main concern is with the research level of the knowledge and philosophy phases and with suggesting abstract "solutions" to implementation personnel. Not concerned in a constructive way with other phases and levels and not integrated into the policy process. Often communicates suggestions with one or two-page statements that are ignored.

Source: Hayden 1993.

Technology

Before beginning to discuss the particular phases and levels in Figure 10-1, to which most of this chapter will be devoted, this section will address the issue of technology. It is important to emphasize that the study of any phase-level conjunction in Figure 10-1 should focus on technology. One of the primary reasons for the failure of modern society to solve its problems is the failure of policymakers and policy scientists to direct their focus in this way. The aspect of policymaking most ignored in the policy science literature is technology. The great power of technology comes from its important role in defining and determining social relationships. Therefore, one of the first ingredients to be considered for any problem is technology. The way we live, the way we relate to each other, the way we communicate, or whether we do, are all heavily influenced by technology. Technology, as defined earlier, is the combination of tools, skills, and knowledge, which are organized as the industrial arts of a society. Its change stimulates creation of new social relationships and thus a new society.

The social sciences have long understood technology's impact on well-being and its impact on the structure and process of society. Institutional economics was built on these ideas. "The institutionalists saw technological innovation as the preeminent factor which determines the institutional superstructure of modern society."[3]

Technology is not the consequence of some benign natural evolution. The tools, knowledge base, and skills are deliberate acquisitions that are usually "ceremonially encapsulated," to use Paul D. Bush's term,[4] by the social forces that have the power and policy means to direct and guide the development and use of technology. Polanyi believed the strength of the policy-technology connection was such that policy, *not process*, determines alternative technology as well as alternative ways of instituting technology.[5] Veblen was pessimistic about the possibility of technology being directed for the good of the social whole because of the encapsulated hold that the corporate business world had on technology. Reminiscent of Veblen, Dewey stated, "[T]he simple fact is that technological industry has not operated with any degree of freedom. It has been confined and deflected at every point; it has never taken its own course. The engineer has worked in subordination to the business manager whose primary concern is not with wealth but with the interest of property. . . ."[6] More recently, the ability of corporations to guide, control, and suppress technology has received considerable attention.[7]

From a policy analysis perspective, real technological advancement or progress, when subjected to a serious transactional assessment, is much more infrequent than usually assumed. To be sure, there is a vast explosion and proliferation of new knowledge, skills, and innovations in the form of new gadgetry, new molecular combinations, computers, engineered genetic mutations, artificial intelligence, and cloning. And these are often combined to provide for human, social, economic, and general biophysical enhancement. Yet too often, the combination is not enhancing, but, rather, deteriorating. Knowledge, intelligence, and inquiry—what we usually call research—are powerful weapons in the determination of the kind and structure of technology that will be instituted and of the enhancing or deteriorating uses to which it will be put. For this reason the research universities and think tanks have become a fierce battleground in the political battle to direct and control research and technology.

The established centers of corporate power have learned the lesson that technology structures society, that it influences the degree of centralization in the decision processes and influences the condition of ecosystems. Corporations have arrived at the conclusion that social and physical technology can be controlled to a great extent by controlling research centers. By influencing the selection of researchers and professors, and guiding the funding for research, they can also influence and guide the kind of technology, society, decision processes, and ecosystems that will emerge. With this realization, business corporations, along with other power centers with a similar interest and ideology, have invaded the universities. The power structure and conditions of life are at stake; thus, universities have become combat zones for their determination. The battle has been fierce, but to date very one-sided with corporate power centers generally prevailing. One example is the success regulated industries have had in promoting the theoretically bankrupt ideas of neoclassical cost-benefit analysis. David Bollier and Joan Claybrook explain this success as follows:

> The success of regulated industries in winning respectability for cost-benefit analysis is symptomatic of an important political fact: industry dominance of regulatory knowledge and debate. By funding public policy institutes, trade associations, research projects, and university chairs, regulated industries have helped

underwrite scholarship stressing the costs and constraints imposed by regulation.

The void in knowledge about health and safety regulation can be traced to the superior resources that industries command in generating regulator knowledge and disseminating it . . . . Victims of corporate misconduct . . . lack the financial means and political organization to give greater intellectual dimension or currency to the beneficial freedoms that regulation can secure for them.[8]

Those economic power systems that stand in opposition to the development and use of technology to enhance human life and the ecosystem have their act together—from theory to bureaucratic appointments, from ideology to computer data systems, from rhetoric to power bases, from the provision of research funds to the sponsorship of scientific journals, from the influencing of universities to the destruction of regulations.   They are organized and they are delivering.   They are effectively wielding complete paradigms, from philosophy and theory to advocacy and operations.

Technology, which is one of the most important ingredients of human welfare, has become a foul word in the minds of many people because it is so regularly associated with hazardous spills, unemployment, cancer, community disruption, consumer victimization, ozone depletion, and so forth.   If technology is to advance in the sense of enhancing progress for human and ecosystem welfare, the people's legislative bodies must explicitly and directly take back control of the research functions of their public universities.

The concept of technological change without progress is not new.   John Dewey recognized that in a holistic sense, advancement is infrequent[9] because the expanded ability to capture energy, increase speed, and process information often is not matched by the ability of human governing bodies to analyze and control themselves and their technology.   But Dewey held "steadfast to his faith in science, in collective intelligence, and in a machine age that . . . will be a means of life and not its despotic master."[10]   For Dewey, "the evils in our system, . . . call for knowledge and scientific insight to surmount difficulties and eliminate sources of defect and ill; they call for a renewal of spirit, for moral development, and for a remaking and redirection of social forces and conditions."[11]   None of these relates to the call now being imposed on research universities by corporate donors.

Policymakers have been concerned with determining, through technological assessment, which direction is forward. But there is no implication "that something ought to be done simply because it can be done, scientifically and technologically. We do not intend the intrusion of any kind of technological determinism."[12]

Metapolicymaking

The remainder of the chapter will be devoted to consideration of the phases and levels of Figure 10-1. The three levels—policy, strategy, and tactics—for each phase will be discussed, beginning with Phase I: Theory, and continuing through Phase X: Sociotechnical Change.

Phase I: Theory

A. Policy: Creation of Integrated Systems Knowledge. The titles for the levels in this phase (Figure 10-1, left-hand column) are taken from the title of the journal, *Knowledge: Creation, Diffusion, and Utilization.* The concept of the "creation of knowledge" is particularly germane to the inquiry and discovery process of policy analysis and to the creation of theories and warranted assertions. This concept recognizes that science and knowledge creation are completed by humans, through discretionary action within their social processes. Knowledge is not "out there' waiting to be discovered. It is created. As Gunnar Myrdal stated, there is an inescapable a priori element in all scientific work. This means the frame of reference is not given for analysts; it is created by them.

As will be explained later, the a priori assumptions and frame of reference for knowledge creation need to be consistent with the context of the problem to be solved; otherwise the resulting theories and warranted assertions will not be relevant to policymaking. Milton Lower explained that this is why Veblen cautioned against the "given and immutable" frame of economists in the classical tradition. Veblen said of that tradition, it "limits their inquiry in a particular and decisive way. It shuts off the inquiry at the point where the modern scientific interest sets in."[13] Therefore, when neoclassical economists insist on the market system ideology as their a priori element and market models as their frame of reference, the findings of their inquiry are usually irrelevant to policymaking. What was left of the "given and immutable" competitive market system that is the base for the neoclassical model was swept away around the globe in the 1930s. In the United States, institutional economists played a major role in structuring the New Deal that designed and implemented the welfare state, consistent with the work of Veblen and Commons. Prominant institutionalists such as Rexford Tugwell and Adolf Berle were major advisors to President Roosevelt. The thrust and direction of the New Deal has continued especially in the environmental protection and citizen entitlement areas.[14]

The "alphabet soup" of the SEC (Securities and Exchange Commission), FDIC (Federal Deposit Insurance Commission), FDA (Food and Drug Administration), EPA (Environmental Protection Agency), ASCSS (Agricultural Stabilization and Conservation Service), and SCS (Soil Conservation Service) begun in the New Deal, or added since, is a core which helps guide our lives and regulates the economy in the United States today. The frame of reference for the creation of knowledge needs to be relevant to this reality.

B. Strategy: Diffusion of Knowledge. "Information processes are part of a policymaking process. Information is one of the main bases for power and influence."[15] Some even depict "professionals as dominating the policy process through professional issue networks in and out of government that have replaced the old, special-interest pressure groups."[16] William Melody has explained how information flows and communication technologies are central to policymaking as well as how the control of these by powerful interests can determine the kind and flow of decisions. "A major challenge for public policy will be to find methods to insure that developments in the information and communication sector do not exacerbate class divisions in society and that its benefits are spread across all classes."[17]

C. Tactics: Utilization of Knowledge. A well-developed scientific base with regard to the utilization of knowledge now exists. From that work, a number of conclusions can be gleaned.

First, multidiscipline interfacing is necessary on a regular basis.

Second, it is necessary for researchers to stay involved in the application process of scientific findings if those findings are not to be corrupted. Research has shown that "information is processed in wondrous ways, few of which are replicative of the original information."[18]

Information is a weapon—a rather powerful weapon—in the policymaking process; thus it is abused and misused. Scientists need to regularly correct misinterpretations of their work and to convey through the policymaking process, which includes the media in a democracy, when their findings are being incorrectly utilized to achieve misdirected policy.

Third, applied research needs to provide for the dovetailing of research information and the information needed by planners and policymakers. The results of policy science research, if they are to be adopted, need to be translated into numerous formats and modes for utilization by those not familiar with the original scientific knowledge. The policy tools discovered will need to be written up for the media; data must be transferred to agency computers; new software must often be developed for use in the public policy research agencies; and training manuals are needed to train operational personnel on the use of new technology. Much of the literature regarding the innovation of basic scientific ideas into commercialized products applies here. Policy science utilization must deal with the same kind of technology transfer problems as other products.

Fourth, it is necessary to deal with the concept of information lead. This concept is in agreement with Melody's ideas presented above: "In order to have power and influence, an actor should build up an information lead over the other actors, or, on the other hand, reduce his information arrear."[19] Or, stated differently, excellent research is of little use if it is not timely.

The need for control in the policymaking process is obvious at the strategic and practical levels when observing Figure 10-1. There is also need for standardization and control at the policy-research level. When verification of theories or social research techniques is sufficient to give them warrant, they are standardized.

There can be no intelligent objection to standardizing instrumental equipment, theoretical and practical. Standardization is necessary for efficiency and precision in control. But there is fundamental cause for intelligent objection when control over the standardized equipment is substituted for control in the solution of an actual problem which the use of the standardized equipment can give. When such substitution is made, the use of the equipment, instead of enriching experience and helping its growth, stunts and distorts it.[20]

Phase II: Philosophy

A. Policy: Instrumentalism. The philosophy recommended, as explained earlier, is instrumentalism, as Dewey named it. It is a philosophy inherently concerned with policy. "For those who believe it is the philosopher's task to juggle the universe on the point of an argument, Dewey is a complete disappointment. The world he starts out with and also ends with is the common world we all live in and experience every day of our lives."[21] As James Street explained, "instrumental valuation was concerned with the intellectual selection of future alternative actions,"[22] or, as William James stated, the instrumentalist method "is to try to interpret each notion by tracing its respective practical consequences."[23]

B. Strategy: Emphasize Problems and Consequences. An orientation toward problems and consequences is the instrumentalist strategy. It is also an effectiveness strategy. Great effort can be saved during research, deliberations, discussions, and lobbying if the focus is maintained on the problem at hand and the need to look at consequences. Alternative agendas can be turned aside if discussions are continually brought back to the problem. If policymakers and analysts continue to emphasize the consequences of a problem—how many children are dying, how much venture capital is needed, the extent of soil erosion—attention will be directed to policy research for achieving altered consequences. People *are* concerned with problems and consequences. Policy advisors should not distract by bringing alternative ideologies and personalities into the discussion. They should bring the knowledge base to bear and tenaciously maintain, in very clinical language, the focus on the problems and consequences. There is power in the problem and consequences strategy—power well beyond political portfolio. The basis of legitimacy of policy scientists is research competence. They cannot appeal to a political base; they are brought into the policymaking process to recommend answers to problems.

When instrumentalists give testimony, irrelevant questions may well be posed. It is wise, for the purposes of informing, to turn every question toward the relevant research base of the problem at hand and to use the time to explain the research. This is true whether in open testimony or back room discussions. Precious clock time available should not be used to explain why the question is irrelevant. The opposition is under the same pressure of clock time so they will probably not want the session to end without making their own points. Only if the opposition severely attacks for not answering the question (now they look like bullies) does one explain why the question as directed is irrelevant. Explanations should be offered with politeness (still in clinical language) and thoroughness (now they look like irrelevant bullies). It is important for the legislators, the press, the bureaucracy, and lobbyists to deal with the problem and consequences as defined by the research. Only with that kind of intense focus will its strengths, weaknesses, and meaning become understood.

Policy analysts are not elected officials. They should not undermine their legitimacy and credibility by making the conjectures that are rightly expected of politicians. They should not become advocates of a view for a politician without a verified research base. If they act like politicians, they will divert attention from problems and consequences. If they want to assume the role of a politician, they should run for election.

The instrumental researcher's strategy should be to focus attention constantly on the set bounded by the problem and its consequences, and the incremental changes needed to transform the first set into the latter set.

C. Tactics: Build Participatory Democracy. None of the policy considerations and strategies discussed above are possible without a democracy that provides for and ensures freedom of inquiry, open hearings, problem-solving processes, public research universities, the diffusion and utilization of information and knowledge, and so forth. Government intervention to sustain such conditions through democratic processes should be relevant and continuous.

Democracy is the tactical necessity and operational expression of instrumentalism. It is also a weapon—the most powerful weapon instrumentalists have. In all respects of policymaking, they should constantly strengthen democracy and its operation. The assistance of political scientists and public administration scholars must be enlisted to help find better ways to make a democracy function at the operational level. Research, philosophy, norms, and legislation can all be lost if the operational tactics are not consistent with the democratic ideal. How agencies should be structured, or procedures routed, or monitoring conducted, or bureaucrats disciplined, or task force membership determined, or administrators selected, or research and development contracts structured—are all issues to be considered at the tactical level in order to strengthen democracy.

Conservative economists such as Milton Friedman and James Buchanan understood that democracy is the enemy of the unfettered market system, or of any other dogmatic ideology that is intended to structure the economy in a manner to make humans and their institutions subservient to it in an immutable manner. Both men called for changes in the system of Western democracy to stop democratic processes from intervening through the government to solve social problems. In Polanyi's terms, they want a system in which the economic and financial institutions dominate, structure, and direct society.

Does this mean that instrumentalists cannot complete policy research or offer relevant advice in nondemocratic societies? No, it means that they can never be as effective as under conditions of democracy. It also means that they should use every opportunity to complete research and seek advice from oversight committees committed to open critique and discussion as in a democracy. Thus in conclusion, instrumentalist tactics are to approach all the phases and levels in Figure 10-1 in a manner consistent with democracy. Progress in that direction is more possible in some cases than in others.

Phase III: Ideology

A. Policy: Community Perspective. Instrumentalism is an ideology. It is a broad base of beliefs about knowledge, philosophy, ceremony, technology, government, and political theory—beliefs that are organized in a systemic and congruent manner. An ideology is the integration and systematization of congruent beliefs. The institutionalist ideology is certainly a different kind of ideology than most, but it does fit the definition of an ideology. It differs in kind from ideologies that prescribe immutable structures and behavior patterns predetermined for societies and humans, respectively. Instrumentalism has no predetermined socioeconomic structure to dictate to people. It is an approach to science, evaluation, and policymaking—an approach that arrives at beliefs through scientific inquiry. These beliefs regarding policymaking were outlined by Jerry Petr. He stated that this "approach to economic policy is (1) values driven, (2) process-oriented, (3) instrumental, (4) evolutionary, (5) activist, (6) fact-based, (7) technologically focused, (8) holistic, (9) non-dogmatic, and (10) democratic."[24] These are beliefs, which are systemic and integrated, beliefs about the most effective way to approach policy. Generally, this ideology can be categorized as community oriented, meaning it is believed that people's lives are organized and their welfare determined by a community's organic social process. The community, which is more than the sum of its parts, has special ongoing needs. If those needs are not regularly met, whether it be factory, neighborhood, city, or global community, the people in the community will be alienated, frustrated, and without means to satisfy basic needs. The promise of policymaking can be fulfilled if there is a sense of societal purpose, "an end of fragmentation and individualism, a coming together around the need for more equitable distribution of wealth and power among all . . . ."[25]   This includes the community's explicit provision of citizen membership rights for goods and services such as education, credit, housing, health care, and so forth.   For a well-designed community that can coordinate the resources necessary to provide the goods and services (without bads and disservices such as excessive degradation of the natural environment and disintegration of the community) necessary to fulfill membership rights, government planning is required.   Instrumentalists "acknowledge that government has the responsibility and should have the capacity to perform the task of community analysis and planning, as well as of determining priorities and allocating resources accordingly."[26]   Such determinations and allocations must, as noted above, be and remain democratically accountable.

B.   Strategy: Utilize Holistic Systems Analysis. The ideological strategy of communitarianism is to practice holism; to take a holistic approach to all aspects of policymaking, whether it is in conducting research, lobbying, building statistical bases, or writing a legislative bill.

Holism is a modeling approach and perception of reality that integrates real-world elements and components into wholes that are comprehensive systems.   Holism rejects the atomistic and reductionist approach.   It has been found that reality is organized so that transactional (rather than interactional or self-actional) wholes guide and determine the behavior of the parts of a system.   This should be heeded throughout the policymaking process.   For example, many databases have been compiled from a reductionist point of view.   They cannot be recycled into holistic models.   Sloppy science, as stated above, harms people and environments.   "The scientist cannot be left to devise our undoing, however unwittingly."[27]   Neither can the planner or policymaker. "The old idea of scientific specialization has given way to a new conscientiousness of the interrelatedness of all things. Spaceship earth, the limits of growth, the fragility of our life-supporting biosphere have all dramatized the ecological and philosophical truth that everything is related to everything else."[28]   It is going to be necessary to educate large numbers of experts in holistic science and planning tools. Without a large cadre of competent holists, we will continue to throw resources down reductionist holes; billions will be provided for large-scale programs that make life less bearable.

C. Tactics: Select Beliefs, Attitudes, Actors, and Criteria. One of the reasons that governments continue to spend billions without constructive or progressive results is that there has not been a detailed concern to make tactics consistent with strategy.

At the tactical level, it is necessary to select specific beliefs and attitudes consistent with the communitarian ideology and to implement them through the selection of relevant criteria and actors.   It is especially necessary to select the correct actors at the level where policy, strategy, and tactics are conducted.   This cannot be done without personnel who are not only competent in their area of expertise but who also have the appropriate beliefs, attitudes, and ideology.   If the minds of operation personnel are not in sympathy with the beliefs about what is to be accomplished, it is unlikely that much will be accomplished. A potpourri of beliefs and ideologies among cabinet officers and ministers, and among directors in the bureaucracies, will lead to a potpourri of de facto policies and programs that will at best be in conflict with each other, and worse, move policy in the wrong direction.

Phase IV: Problem Definition

A.   Policy: Structure and Process Oriented Description.   As stated earlier, the problem orientation is the only reliable way that instrumentalists have found to organize scientific research and policymaking. Since solving social problems always requires changing institutions, and in many cases technology, the problem should be described in terms of the structure and process that are delivering the problem. With most studies, "the examination of the problem is all too often neglected and not continually repeated throughout the course of the study. Thus the result is that many fine efforts are directed at a problem which is not at the heart of the matter."[29] Social problems, from the delivery of carcinogens to the delivery of low incomes, are delivered through the process of the social structure. Policy and process are inseparable. The goal is to define the sequence of events in a process. Thus, problem definition should be a structure and process description.

B. Strategy: Define Sociotechnical and Socioecological Setting.
The context of any research is important. If the context is not relevant to the problem to be solved, the structure and process that need to be impacted by policy will not be the appropriate ones. Therefore, the problem definition should be embedded in the sociotechnical setting in order to know the institutions and technologies that need to be changed. This means a great deal of the research should be done in the field, standard operating procedure manuals, and court records.

C. Tactics: Create Narrative and Statistical File. As noted, the problem is too often neglected in policy research. The author's experience has found the same to be true at the tactical level. Managers, administrators, and fiscal analysts seldom are aware of the research that initiated programs, how the problem was defined, or the context of the problem. Thus, the definitions and findings, including notes from field observations and surveys, should be brought on line in the computer system for ready access in the bureaucracies. The problem definition should be readily available to tacticians in all phases so they are constantly reminded of what problems they are attempting to solve.

Phase V: Context

A. Policy: Conceptual Framework of the Social Fabric Matrix and Digraph. Harold Lasswell, whose work is often credited with creating the area recognized as policy science, emphasized, consistent with what has been explained above, that the approach of policy scientists is problem-oriented and contextual. A framework is needed to insure that the context is holistic and transactional. Most policy scientists agree that the main deficiency in the policy sciences is the lack of an integrated framework to carry the theory and research into the policy area in an organized and effective manner. Yngve Ramstad has stated that theory "will be meaningful (instrumental) only if it permits one to *act* correctly, that is to show how institutions can be altered *in a spe-*

cific context so as to actually effect the intended consequences."[30] He added that the social fabric matrix and digraph approach is the holistic framework best suited for such an instrumental endeavor. The SFM allows for diverse technical expertise to be harnessed into a unified system to strengthen evaluation and decision making. Thus, the character and structure of the SFM is an instrumental tool for organizing policy analysis for complex systems.

B.   Strategy: Case Study Approach. The strategy of the SFM approach was explained above in Chapters 6 and 7. As is evident from the description above, the context strategy of instrumental policymaking is case-by-case studies that are problem-oriented and can be completed with the holistic SFM framework.

C.   Tactics: Computerize Spreadsheets and Statistical Analysis. The social fabric matrix research can be made operative at the tactical level by computerizing the digraph spreadsheets as explained above, and thereby be utilized for agency monitoring, data updating, and statistical analysis.

Phase VI: Measurement

A.   Policy: Contextual Social Indicators. The social indicator movement that began in the 1960s is no longer a movement. Now it is understood that all useful measures are social. It is now broadly understood that price is not *the* measure of either programs or benefits, nor is there any other measure that can be a common denominator. Various interests have promoted various elixirs that were to serve the common denominator function.   There is no such easy solution to measurement.   Indicators must be developed consistent with the problem, the context, and the ideological criteria. "There is the recognition that all observation, all measurement, all experience is necessarily subjective.   Neither the measurer, the measure, nor the measured is absolute."[31]   The policy with regard to indicators was explained above in Chapter 5.

B.   Strategy: Determine Consequences, Requirements, Relationships, and Monitoring Indicators. Following Dewey's advice, the strategy with regard to social indicators for policymaking is to derive the indicators and database needed to complete the contextual analysis with the social fabric matrix and to monitor policy. As was explained in Chapter 5, it is necessary to develop indicators that can be used to indicate the following: 1) consequences, 2) requirements, 3) relationships, and 4) monitoring. All of these are defined and available in the social fabric matrix and digraph.

C. Tactics: Construction of Databases. These measurement indicators should be established in the governmental computer mainframe and made readily accessible as a database for agencies, lobbyists, and citizens. As tacticians from those groups insist on data that are useful in making policy decisions, we can overcome one of the lingering criticisms of social indicators. Possibly the most striking error of commission for which the societal accounting movement can be faulted has been the lack of decision making relevance of its products.

The irony of the so-called "Information Age" is that information is less and less available for two reasons. First, although there is more and more concern with the latest hardware and software, there is less and less concern about the theoretical and conceptual base for what is useful and what is useless information. The volume of data could be drastically cut if the data were meaningfully organized. It is not so organized; therefore, many users have great volumes of data but little useful information. Second, many people can no longer access the information sources. In the past, printed copy was available to almost everyone through public and university libraries. Now that more and more of our databases and text are available only via computer access, fewer citizens can access the information base necessary for democratic deliberations and control.

The SFM and measurement approach explained above provides a conceptual framework for meaningful information and knowledge. The tactical responsibility is to provide computer systems so that all agencies can access each other's databases and provide computer access to citizens as well.

Phase VII: Select Programs

A. Policy: Design Alternative Programs. There is no training or framework with which the author is acquainted that guarantees that policy scientists can arrive at creative program alternatives. The policy scientist will be in a much better position to be insightful and design viable alternative programs if he or she has been immersed in Phases I through VI. Of course, neither should a bit of experience at the strategy or tactical levels in Phases VII through X hinder (although it sometimes does). A mixture of scientific research, field work, policy experience, and a review of solutions utilized in other societies is usually helpful but not always sufficient to design creative policy alternatives.

B. Strategy: Test Alternatives in the Social Fabric Matrix and Digraph. The strategy for program development is to test alternative programs with the social fabric matrix to determine direct and indirect consequences.   The SFM articulates and describes the structure, process, and deliveries of the problem area.   Programs designed to solve the problem can be tested in the SFM.   Selection criteria can be applied in order to determine if the program created has improved or deteriorated the situation.

C.   Tactics: Efficiency Tests Based on Consequences and Effectiveness.   The tactical task is to judge the myriad consequences that any program manifests and to select the best program.   Because there is no common denominator, policymakers have the opportunity to judge the different mixes of results projected for the different programs.

Effectiveness can be determined for the different programs by comparing the total consequences of each program with the budgetary requirements of each program.   Therefore, it provides an alternative to the cost-benefit analysis of neoclassical economics.

Phase VIII: Advocacy

A. Policy: Advocacy Research. Policy advocacy is a matter of doing the research, getting organized, and working your network.   Sound easy?   It is not.   It is the most difficult, with the least probability of success (although often great fun), of all the phases.   All the resistance to change, to include ceremonial lag as well as established power, comes into play.   Advocacy research integrates, combines, and repackages basic research in order to persuade.   One of the best procedures for pursuing advocacy research is the use of a citizen task force if the task force is provided with an adequate and competent staff that can utilize the scientific research base already completed.

Because advocacy research builds on basic policy research, there sometimes is a communication problem. First, as stated earlier, a tremendous knowledge gap exists in the transfer process due to the lack of attention to the diffusion and utilization of knowledge. Advocates may not be able to understand the basic research, and the scientists may have no contact with the advocates. Second, the interest of advocates is to pursue the interest of their group and therefore they have a tendency to be less concerned than they should be about the quality of the basic research that they utilize to support those interests. The third and most severe problem is that there is a paucity of research being conducted that provides adequate understanding of modern problems. Before basic research can be utilized, it must already have been completed. President Lyndon Johnson, when asked what is the most important ingredient to policy success, is purported to have said, "Seize the moment! Seize the [expletive deleted] moment!" Relevant basic research needs to be on the shelf when the opportunity for its use presents itself or it will not be possible for advocates to seize the moment to pursue problem resolution through policy initiatives.

This chapter is not about politics; it is not a handbook on how to run a political campaign. Advocacy, however, comes close to political campaigning, and shares some of the same goals and techniques. This means that a special effort must be made not to allow policy research to be subverted by politics. However, a couple of caveats concerning any presumed neat separation of advocacy and politics are in order, lest we think we can become sanitized from politics. First, a policy or idea entrepreneur without some political savvy is just another promoter. Second, let's remember how John R. Commons explained the commonality between politicians and others involved in concerted effort. He wrote, "political parties, like all concerted action, . . . have the very practical purpose of getting and keeping control of the officials who formulate the will of the state."[32] In the same way, those engaged in policy or advocacy research cannot ignore politics. Neither should advocates ignore that they are in a power struggle. Concerted action is "designed to get and keep control of the concern and its participants."[33] Nor should we be surprised when opponents who are in that struggle flex their muscles or kick below the belt. Be assured they do. This is why progress is slow at best, and extremely difficult in the usual case. In a year, "there will be maybe 11,000 bills introduced in this Congress. Maybe 600 or 650 will be passed, and out of those, 400 to 500 are standard reauthorization bills. So what we are really taking about is a salmon run of policy and ideas. You've got to have persistence. When you're working ideas, you've got to be prepared to hang in there."[34]   In addition to the counsel for persistence, I would add organization.

B.   Strategy: Organize Elements and Symbols. The strategy for advocacy is organization, organization, and more organization.   The computer is now key to that organization.   It is no longer possible to match the opposition without computer-based organization.   Policy and program advocacy is where task force reports, corporate jets, favors, lobbyists, scientific testimony, media manipulation, special interests, direct mail pieces, and computers are mixed—hopefully in a fine-tuned manner—and usually fail, especially in the short term.   Every aspect of the advocacy process has to be organized: the corporation jets to take advocates to Washington, DC, Austin, or Paris; the production and mailing of the four-page direct-mail piece; the identification and briefing of those who will give public testimony; the letters-to-the-editor campaign; the schedules and volunteers; the free media effort; the paid media effort; the areas to be canvassed door to door; money for the canvassers.   They all need to be organized.   On and on organization grows.

Strategy studies for organizing policy advocacy have received little attention from instrumentalists or from academics in general, although *advocacy is an important linkage between policy research and the adoption and implementation of new programs.* Even though social theory has dealt little with these organizational techniques, it does clarify what must be organized and what must be included in the network and lobbying effort.   Social research defines and explains the elements, beliefs, and ceremonial symbols central to organizing advocacy efforts.

C.   Tactics: Networking and Lobbying. The job of the tacticians and operations personnel is twofold.   First, it is to use symbolic means to achieve instrumental results; to use symbols to acquire substance.   It is the job of the lobbying and networking process to close the gap between symbols and substance by making the consequences wanted into the symbols that others want.   Second, it is to create new symbols when necessary.   Instrumentalists have explained the importance of mental habits, the power of symbols, and the relevance of both in patterning social institutions for both instrumental and deleterious purposes.

The life process, to include policymaking, takes place in a social and cultural milieu in which symbols are essential for organization and communication. The anthropologist, Raymond Firth, has stated that "man does not live by symbols alone, but man orders and interprets his reality by his symbols and even reconstructs it."[35] Thus, as Charles Elder and Roger Cobb explain: "Public policymaking tends to be a highly stylized and ritualized process. It is replete with symbolism that conveys reassurances and serves to rationalize the product, whatever it may be."[36] Policy innovations need social and cultural moorings.

A social symbol is a human invention; "people invent them, acquire them by learning, adapt them, use them for their own purposes."[37] New symbols are "likely to occur in the face of dramatic events or major changes in the natural, social, or political environment."[38] Symbols may "emerge as a consequence of their deliberate advocacy by political leaders or issue entrepreneurs."[39] In some cases, a new policy may be so different from traditional institutional norms that new symbols will need to be generated to adequately represent the new policy. "New symbols can be created and old ones redefined (or discredited) so as to create a climate conducive to a significant policy innovation."[40]

A significant difference needs to be noted between traditional lobbying and the standard for instrumentalists. Traditional lobbying has too often appealed to, and thereby reinforced, any symbol, whether or not the symbol and the institutions it represented were deleterious. Consistent with Dewey, means determine ends, so if deleterious symbol manipulation is used, it will encourage a deleterious result. Events must be exploited through effective symbol management. The instrumentalist standard, however, is that symbols should not be utilized that will reinforce deleterious consequences or institutions. Symbols must be utilized to generate human responses. Lobbying, to be consistent with instrumentalist thought, requires that social symbols and institutions to which the lobbying effort is appealing be instrumental. In this way, lobbying for a new program is also reinforcing social and mental habits that have been found to be instrumental.

In addition, an education and propaganda process will need to be developed in order to convince people of new symbols and social patterns needed to make the program a success.

Lobbying is referred to as the fifth estate of government, although not necessarily the fifth in power. There is more than the vast inequality in access to lobbying funds that is fueling the growth in lobbying. The more technology grows, and consequently, the more society becomes differentiated and complex, the more the lobbying component must and will grow. Lobbying will grow, and lobbyists, who are becoming much more technically competent about their subject matter, are crucial to policymaking. It is no compliment to most universities that they have masters' degrees in business administration but not the equivalent for educating lobbyists.

Instrumental research is ready-made for lobbying and organizing networks. It contains both the skeleton and the flesh. "The concerted action of politics within a concern is founded on passion, stupidity, inequality and mass action, yet it can be investigated scientifically. . . ."[41] Such scientific inquiry is that to which instrumental research is devoted. It identifies power bases, deliveries, losers, winners, social groups, key social actors, institutions, beliefs, inequalities, government agencies, corporate interlocks, and so forth. Before an effective lobbying effort can be mounted, the actors, institutions, technologies, and financial flows must be identified. Instrumental analysis is problem oriented; thus, it turns to the analysis needed to analyze the problem. As problems change, different methods, principles, data, and statistical techniques are needed. As problems change, different alliances are needed for policymaking. The analysis of the problem tells us with whom the lobbying effort should seek alliances. It will be bankers in some cases, the GI Forum in others, the Sierra Club in others; and all three together in other cases. As the problem changes, the alliances change. Lobbyists are familiar with this. This is why civility and honesty are important among policy enemies. Today's enemy is tomorrow's friend in policymaking.

Phase IX: Program Budgeting

A. Policy: Monetary, Rules, and Judicial Budgeting. Budgeting for programs is the process of allocating resources to continue old programs and create new ones. It includes: 1) the allocation of money flows, 2) making statutory changes, 3) making administrative changes, and 4) bringing about new judicial rulings in order to create new institutions, new technologies, and new human behavior patterns. Budgeting is not just a matter of allocating money. Probably far more resources are allocated and institutions changed through statutory, administrative, regulatory, and judicial means than through monetary allocation. Taxation, bond indebtedness, and monetary allocation theories are well developed. What needs research attention is the design of budgeting systems that reflect budgeting beyond the allocation of money.

B. Strategy: Agency Allocation. The program budgeting strategy is to allocate and coordinate the monetary, statutory, administrative, and judicial changes across the agencies in a manner to allow for the creation of new instrumentalities necessary for delivering social beliefs and social programs to the correct clients.

C. Tactics: Budget Requests and Rule Changes. The accountants, auditors, and fiscal analysts need to have skills and knowledge from an array of scientific disciplines to carry out the budget requests and performance audits. That has been understood by many budget agencies that have hired multidisciplinary staffs. However, the multidisciplinary personnel many times continue to be forced into the mold of narrow models and dualistic double-entry accounting systems. As stated earlier, these are the actors most likely to become divorced from the original values, beliefs, and intent of programs. The tactical recommendation for all the phases has been designed, in part, to try to overcome this problem

Phase X: Sociotechnical Change

A. Policy: Creation of Institutions. The final phase is the building of new institutions consistent with programs and budgets. At the policy science level the main research activity is the design of institutions, the definition of roles to make the institutions effective, and the forecasting of behavior-effective models.

B. Strategy: Creation of Instrumentalities. The strategic level of sociotechnical change is concerned with the creation of instrumentalities. The emphasis is on specific programs that need to be accomplished and strategies to make them a reality. These are usually applied systems that are manifest mainly in social and technological innovations; new airport systems, new disease prevention systems, new weapon systems, new industry regulation systems, new environmental protection systems, and so forth.

C. Tactics: Operations and Procedures. At the tactical level are the technicians, clerks, word processors, labor unions, tractors, knowledge bases, water supplies, and so forth that are to be managed and administered in efficient systems. This level actually puts the system variables in the right place each day to deliver the results of the policymaking process. Theory with regard to the rigidity of mental habits is very relevant here. Old-line managers and bureaucrats with skills, knowledge, tools, and ideology from an undergraduate or graduate program of the distant past are often unable to understand the connection between belief systems and their own operations. Many times those who have worked their way up through the bureaucracy have learned "the system," and the last thing they want to do is endanger their positions with a new mode of operation. It is important to realize that if operations and procedures in the active bureaucracy are inconsistent with the theory, intent, and strategy of policymaking, then pushing progressive bills through legislative bodies or achieving instrumental court decisions will be of little value. Likewise, to design policy research or programs without considering the viability of tactical operations to finalize the research findings and programs, limits the value of the research no matter how intellectually pleasing or elegant it may be. "Instrumentalists see planning and administrative 'control' as essential ingredients of economic organization whether in the public or the private spheres."[42]

Summary

In summary, the emphasis has been to extend and broaden the tool kit of instrumental policymaking to include research and expertise for all the phase-level conjunctions of Figure 10-1. Without this extended base, the delivery of instrumental policy can break down. Policy science research cannot be utilized without effective principles from the strategic, management, and administrative sciences that have been grounded in a similar intellectual tradition. Likewise, managers and administrators are doing little more than carrying out procedures if they are not operating consistent with the philosophy, ideology, and research from the policy level, as is indicated in Figures 10-2, 10-3, and 10-4. Relevant policymaking research on any complex social problem requires expertise in all phases of policy research. Significant policymaking for a good society requires concerted efforts to coordinate policy research with all phases of strategy and tactics.

NOTES AND REFERENCES

Chapter 1: Introduction

1. The emphasis on the importance of analytical means is not intended to convey a perception that numerous other aspects are not involved in preventing good policy. My experience in the policymaking arena precludes such naivete. Experiences have made me painfully aware, for example, of the tremendous power and influence of entrenched corporate interests in the policy process. That experience has also made me aware that one of the strongest weapons against such interests is well publicized results from competent research. For an example of the power of research and advocacy with regard to reducing the power of the atomic energy industry in Washington, D.C. see: Duffy, Robert J. *Nuclear Politics in America: A History and Theory of Government Regulations.* Lawrence: University of Kansas Press, 1997.

2. Anderson, Frederick R., Robert L. Glicksman, Daniel R. Mandelker, and A. Dan Tarlock. *Environmental Protection: Law and Policy.* New York: Aspen Law and Business, 1999, p. xxvii.

3. Ibid., p. xxviii.

4. Menand, Louis. *The Metaphysical Club.* New York: Farrar, Straus and Giroux, 2001.

5. Lodge, George C. *The New American Ideology.* New York: Alfred A. Knopf, 1976, p. 3.

6. Menand. Op. cit., p. 441.

7. Westbrook, Robert B. *John Dewey and American Democracy.* Ithaca: Cornell University Press, 1991, pp. 408–409.

8. Lippmann, Walter. *The Good Society.* Boston: Little, Brown, and Company 1937.

9. Bellah, Robert N., Richard Madsen, William M. Sullivan, Ann Swidler, and Steven M. Tipton. *The Good Society.* New York: Alfred A. Knopf, 1991.

10. Daly, Herman E. and John B. Cobb, Jr. *For the Common Good.* Boston: Beacon Press, 1994.

11. Dewey, John. *The Public and Its Problems.* New York: Henry Holt and Company, 1927.

12. Polanyi, Karl. *The Great Transformation.* Boston: Beacon Press, 1944.

13. De Greene, Keynon B. *Sociotechnical Systems.* Englewood Cliffs: Prentice-Hall, Inc., 1973.

14. Lodge. Op. cit.

15. Tool, Marc R. *The Discretionary Economy: A Normative Theory of Political Economy.* Santa Monica, CA: Goodyear Publishing Company, Inc., 1979.

Chapter 2:   Policy Paradigms Should Be Consistent with the Complexity of Reality

1. See Hayden, F. Gregory. "Policy Concerns Regarding Ecologically Sound Disposal of Industrial Waste Materials." In Marc R. Tool and Paul D. Bush, eds. *Institutional Analysis and Economic Policy*. Boston: Kluwer Academic Publishers, 2003, pp. 466-478.

2. Baudrillard, Jean. *The System of Objects*. New York: Verso, 1996, pp. 195-196.

3. English, Mary. "Siting, Justice, and Conceptions of the Good. *Public Affairs Quarterly* 5 (January 1991), p. 9.

4. Young, Iris M. "Justice and Hazardous Waste." In Michael Bradie, Thomas W. Attig, and Nicholas Rescher, eds. *The Applied Turn in Contemporary Philosophy*, pp. 171-183. Bowling Green, OH: Bowling Green Studies in Applied Philosophy, 1983, p. 177.

5. Anderson, Frederick R., Daniel R. Mandelker, and A. Dan Tarlock. *Environmental Protection: Law and Policy*. Boston: Little, Brown, and Company, 1984, p. 35.

6. Dror, Yehezkel. *Public Policymaking Reexamined*. Scranton: Chandler Publishing Co., 1968; and Dror, Yehezkel. *Policymaking Under Adversity*. New Brunswick: Transaction Books, 1986.

7. Dewey, John. *Logic: The Theory of Inquiry*. New York: Henry Holt and Company, 1938.

8. Hamilton, David. "Comment Provoked by Mason's 'Duesenberry Contribution to Consumer Theory.'" *Journal of Economic Issues* 35 (September 2001), p. 746.

Chapter 3:   Instrumental Philosophy and Criteria

1. Anderson, James A. and Elaine E. Englehardt. *The Organizational Self and Ethical Conduct*. Fort Worth: Harcourt College Publishers, 2001, pp. 6-12.

2. Ibid., p. 9.

3. The terminology of self-action, interaction, and transaction is taken from: Handy, Rollo. *Value Theory and the Behavioral Sciences*. Springfield, IL: Charles C. Thomas, 1969, pp. 52-68.

4. Equilibrium models are inconsistent with the open systems of the real world (see Chapter 4).

5. Handy. Op. cit., pp. 59-60.

6. Ibid., p. 60.

7. Geertz, Clifford. "The Impact of the Concept of Culture on the Concept of Man." In Clifford Geertz, ed. *Interpretation of Cultures*. New York: Basic Books, 1973, p. 34.

8. Birkland, Thomas A. *An Introduction to the Policy Process: Theories, Concepts, and Models of Public Policymaking*. New York: M.E. Sharpe, 2001, p. 122.

9. Ibid., p. 123.

10. Westbrook, Robert B. *John Dewey and American Democracy.* Ithaca: Cornell University Press, 1991, p. 245.

11. Ayres, Clarence E. *The Theory of Economic Progress.* New York: Schocken, 1962.

12. Schumpeter, Joseph A. *The Theory of Economic Depvelopment.* New York: Oxford University Press, 1961.

13. Ayres. Op. cit.

14. Ellul, Jacques. *The Technological Society.* New York: Vintage, 1964.

15. Schumacher, E.F. *Small is Beautiful: Economics as if People Mattered.* New York: Harper and Row, 1975.

16. Dewey, John. *The Public and Its Problems,* 2nd ed. (Chicago: The Swallow Press, 1954, p. 108.

17. Ibid.

18. Eco, Umberto. *A Theory of Semiotics.* Bloomington: Indiana University Press, 1979, p. 61.

19. Ibid., p. 66.

20. Okrent, Mark. "The Truth, the Whole Truth, and Nothing but the Truth." *Inquiry* 36 (December 1993), p. 392.

21. Schlagel, Richard H. *Contextual Realism: A Metaphysical Framework for Modern Science.* New York: Paragon House Publishers, 1986, p. 236.

22. Dewey, John. *Theory of Valuation.* Chicago: University of Chicago Press, 1939, p. 49.

23. Pankratz, David B. *Multiculturalism and Public Arts Policy.* Westport, CT: Bergin & Garvey, 1993, p. 22.

24. *The Economist.* "Let Them Eat Pollution." 322 (February 8, 1993), p. 66.

25. *Weisbrod,* Burton A. "Preventing High School Dropouts." In Robert Dorfman, ed. *Measuring Benefits of Government.* Washington, DC: The Brookings Institution, 1965.

26. Anderson, Charles W. *Pragmatic Liberalism.* Chicago: The University of Chicago Press, 1990, pp. 43-44.

27. Schagel. op cit., p. 232.

28. Ibid.

29. Tool, Marc R. *The Discretionary Economy: A Normative Theory of Political Economy.* Santa Monica, CA: Goodyear Publishing Company, Inc., 1979, p. 289.

30. Sadler, D. Royce. "The Origins and Functions of Evaluative Criteria." *Education Theory* 35 (Summer 1985), p. 282.

31. Ibid., p. 292.

32. Olson, Paul A. *The Journey to Wisdom*. Lincoln: University of Nebraska Press, 1995, pp. 1-27.

33. Chisolm, Roderick M. "Reply to Amico on the Problems of the Criterion." *Philosophical Papers* 17 (March 1988), p. 233.

34. Schlagel. Op. cit. p. 236.

35. Ostrom, Eleanor. *Crafting Institutions for Self-Governing Irrigation Systems*. San Francisco: TCS Press, 1992, p. 45.

36. Ibid., p. 47.

37. Schlagel. op. cit., p. 203.

38. McMullin, Ernan. "Two Faces of Science." *Review of Metaphysics* 27 (June 1974), p. 658.

39. Ibid., p. 669.

40. Lee, Donald S. "Adequacy in World Hypotheses: Reconstructing Peppers Criteria." *Metaphilosophy* 14 (April 1983), p. 515.

41. McMullin. Op cit., p. 671.

42. Ibid., p. 675.

43. George, Kathryn Paxton. "Sustainability and the Moral Community." *Agricuture and Human Values* 9 (Fall 1992), pp. 48-57.

44. Dewey, John. *The Quest for Certainty*. New York: G.P. Putnam's Sons, 1929, p. 85.

Chapter 4: General Systems Principles for Policy Analysis

1. Argyal, Andras. *Foundations for a Science of Personality*. Cambridge: Harvard University Press, 1941.

2. Katz, Daniel and Robert L. Kahn. "Common Characteristics of Open Systems." In F.E. Emery, ed. *Systems Thinking*. Baltimore: Penguin Books, Inc., 1976, p. 90.

3. Hall, A.D. and R.E. Fagen. "Systems Organization and the Logic of Relations." In Walter Buckley, ed. *Modern Systems Research for the Behavioral Scientist*. Chicago: Aldine Publishing Co., 1968, p. 81.

4. Ibid., p. 82.

5. De Greene, Kenyon B. *Sociotechnical Systems: Factors in Analysis, Design, and Management*. Englewood Cliffs: Princtice–Hall, Inc., 1973, p. 4.

6. Ibid., p. 36.

7. Katz and Kahn. Op. cit., 101.

8. Rosen, Robert. "Some Systems Theoretics Problems in Biology." In Ervin. Laszlo, ed. *The Relevance of General Systems Theory.* New York: George Braziller, 1972, p. 53.

9. Swaney, James. "Elements of a Neoinstitutional Environmental Economics." *Journal of Economic Issues* 21 (December 1987).

10. Ibid., p. 337.

11. Rosen. Op. cit., p. 53.

12. Ibid.

13. De Greene. Op. cit., p. 37.

14. Katz and Kahn. Op. cit., p. 100.

15. Ackoff, Russell L. "Towards a System of Systems Concepts." *Management Science* 17 (July 1971), p. 670.

16. Pattee, Howard H. "The Role of Instabilities in the Evolution of Control Hierarchies." In Tom R. Burns and Walter Buckley, eds. *Power and Control: Social Structures and Their Transformation.* London: Sage Publications, 1976, p. 179.

17. Pattee, Howard H. *Hierarchy Theory: The Challenge of Complex Systems.* New York: George Braziller, 1973, p. 77.

18. De Greene. Op. cit., p. 47.

19. Ibid.

20. Laszlo. Op cit., p. 19.

21. Pattee. Op. cit., p. 77.

22. Bryant, James W. "Flow Models for Assessing Human Activities." *European Journal of Operations Research* 4 (June 1980), p. 73.

23. De Greene. Op. cit., p. 22.

24. Katz and Kahn. Op. cit. p. 95.

25. De Greene. Op. cit., p. 78.

26. Ibid., p. 7.

27. Pattee, Howard H. "The Complementarity Principle in Biological and Social Structures." *Journal of Social and Biological Structures* 1 (June 1978), p. 99.

28. Katz and Kahn. Op. cit., p. 99.

29. Hunter, David E. and Phillips Whitten. *The Study of Cultural Anthropology.* New York: Harper and Row, 1978, p. 287.

30. Bertalanffy, Ludwig Von. *General Systems Theory: Foundation, Development, Applications.* New York: George Braziller, 1969, p. 226.

31. Ibid., p. 229.

32. Ibid., p. 236.

33. Hall and Fagen. Op. cit., p. 92.

Chapter 5:   Social Criteria and Socioecological Indicators

1. Blumer, Martin. *The Uses of Social Research Investigation in Public Policy-Making.* London: George Allen & Urwin, 1982, p. 51.

2. Veblen, Thorstein B.   *The Theory of the Leisure Class.* New York: The Macmillan Company, 1899, p. 9.

3. Land, Kenneth C. "On the Definition of Social Indicators." *American Sociologist* 6 (November 1970), p. 323.

4. Ibid.

5. Dewey, John. *Logic: The Theory of Inquiry.* New York: Henry Holt and Company, 1938, pp. 200, 205, 211, and 499.

6. McGill, Dan M., ed. *Social Investing.* Homewood, IL: Richard D. Irwin, 1984, pp. 3-4.

7. McKean, Roland N. *Efficiency in Government Through Systems Analysis.* New York: John Wiley & Sons, Inc., 1967, p.25.

8. *State of Ohio v. U.S. Department of Interior*, 880 F.2d 422 (D.C. Cir. 1989).

9. Blumer. Op cit., p. 52.

10. Rohrlich, George F., ed. *Environmental Management.* Cambridge: Balinger Publishing Co., 1976, p. 274.

Chapter 6:   The Social Fabric Matrix

1. Warfield, John N. *Societal Systems: Planning, Policy and Complexity.* New York: John Wiley & Sons, 1976, p. 63.

2. Geertz, Clifford.   "The Impact of the Concept of Culture on the Concept of Man." In Clifford Geertz, ed. *The Interpretation of Cultures.* New York: Basic Books, 1973, p. 17.

3. De Greene, Kenyon B. *Sociotechnical Systems: Factors in Analysis, Design, and Management.* Englewood Cliffs: Prentice–Hall, 1973, p. 3.

4. Rohrlich, George F., ed. *Environmental Management.* Cambridge: Ballinger Publishing Co., 1976, p. 276.

5. Huntington, Samuel P. *The Clash of Civilizations: Remaking the World Order.* New York: Touchstone, 1997.

6. Neale, Walter C. "Institutions." *Journal of Economic Issues* 21 (September 1987), p. 1182.

7. Ibid., p. 1183.

8. Von Wright, Georg H. *Practical Reason.* Oxford: Basil Blackwell, 1983, p. 74.

9. Allport, Gordon W. "The Historical Background of Social Psychology." In Gardner Lindzey and Elliot Aronson, eds. *Handbook of Social Psychology, Volume I: Theory and Method,* pp. 1-46. New York: Random House, 1985, p. 31.

10. Wolman, Benjamin B., ed. *Dictionary of Behavioral Science.* New York: Van Nostrand Reinhold, 1973, p. 243.

11. Allport. Op. cit., p. 36.

12. Ibid., p. 6.

13. Harre, Rom and Roger Lamb, eds. In *The Encyclopedic Dictionary of Psychology.* Cambridge: MIT Press, 1983, p. 591.

14. Allport. Op cit., p. 37.

15. McGuire, William J. "Attitudes and Attitude Change." In Gardner Lindsey and Elliot Aronson, eds. *Handbook of Social Psychology. Volume II: Special Fields and Applications.,* pp. 233-346. New York: Random House, 1985, p. 255.

16. Rokeach, Milton. "Some Unresolved Issues in Theories of Beliefs, Attitudes, and Values." In *Nebraska Symposium on Motivation 1979,* pp. 261-304. Lincoln: University of Nebraska Press, 1980, p. 275.

17. Wright, Will. *Six Guns and Society: A Structural Study of the Western.* Berkeley: University of California Press, 1975, p. 135.

18. Rokeach, Milton. *Beliefs, Attitudes, and Values: A Theory of Organization and Change.* San Francisco: Josey-Bass, Inc., 1968, p. 116.

19. Hamilton, David. "Comment Provoked by Mason's 'Duesenberry's Contribution to Consumer Theory.'" *Journal of Economic Issues* 35 (June 2001), p. 476.

20. Polanyi, Karl. "The Economy as Instituted Process." In Karl Polanyi et al., eds. *Trade and Markets in Early Empires.* Glencoe, IL: Free Press, 1957, p. 250.

21. Elsner, Wolfram. "An Industrial Policy Agenda 2000 and Beyond—Experience, Theory, and Policy." In Wolfram Elsner and John Groenewegen, eds. *Industrial Policies After 2000.* Boston: Kluwer Academic Publishers, 2000, p. 448.

22. Ibid., p. 451.

23. Parkes, Don and W.D. Wallis. 1980. "Graph Theory and the Study of Activity Structure." In Tommy Carlstein, Don Parkes, and Nigel Thrift. *Human Activity and Geography.* New York: John Wiley & Sons, 1980, p. 77.

24. W.E. Moore. quoted in Ibid., p. 76.

25. Mattessich, Richard. *Instrumental Reasoning and Systems Methodology: An Epistemology of the Applied and Social Sciences.* Dordrecht: D. Reidel Publishing, 1978, p. 290.

26. Ibid. p. 289.

27. *State of Ohio v. U.S. Department of Interior.* 880 F. 2d 432. (D.C. Cir. 1989).

Chapter 7: Illustrations of the Social Fabric Matrix

1. High Performance Systems, Inc. (HPS). *An Introduction to Systems Thinking: ithink Software.* Hanover, NH: High Performance Systems, Inc., 1997.

2. Hayden, F. Gregory and Steven R. Bolduc. "Contracts and Costs in a Corporate/Government System Dynamics Network: A United States Case." In Wolfram Elsner and John Groenwegen, eds. *Industrial Policies After 2000.* Boston and Dordrecht: Kluwer Academic Publishers, 2000.

3. Copies of the contracts may be requested by email from either the author, ghayden1@unl.edu, or the Central Interstate Compact Commission, assist@cillrwcc.org.

4. U.S. Department of Energy/Nuclear Regulatory Commission. 10 Code of Federal Regulations §61. Online. Available from http://www.gpoacess.gov/cfr/index.html.

5. Hayden and Bolduc. Op. cit., pp. 250, 252, and 253.

6. Fullwiler, Scott. "A Framework for Analyzing the Daily Federal Funds Market." doctoral dissertation. University of Nebraska, Lincoln, NE, 2001.

7. Ibid., p. 103.

8. Ibid.

9. Ibid.

10. Ibid.

11. Ibid., p. 104.

12. Ibid.

13.

14. Ibid.

15. Ibid.

16. Ibid., pp. 104-105.

17. Ibid., p. 105.

18. Ibid., p. 255.

19. Ibid.

19. Ibid.

20. Natarajan-Marsh, Tara. "Confronting Seasonality: Socioeconomic Analysis of Rural Poverty and Livelihood Strategies in a Dry Land Village." doctoral dissertation. University of Nebraska, Lincoln, NE, 2001.

21. Yang, Youngseok. "Crafting Hierarchical Institutions for Surface Water Resource Management of the Platte River: A Case Study for the Assessment of Institutional Performance and Transformation." doctoral dissertation. University of Nebraska, Lincoln, NE, 1996.

22. For explanations of these laws, see: Ibid., pp. 136-256.

23. Ibid., pp. 241-242.

24. Ibid., pp. 242 and 244.

25. Ibid., pp. 244 and 246.

Chapter 8: Timeliness as the Appropriate Concept of Time

1. Hampden-Turner, Charles and Alfons Trompenaars. *The Seven Cultures of Capitalism.* NY: Doubleday, 1993.

2. Commons, John R. *Legal Foundations of Capitalism.* Madison: University of Wisconsin Press, 1968 [1923], p. 379.

3. Cottle, Thomas J. and Stephen L. Kleinberg. *The Present of Things Future: Explorations of Time in Human Experience .* New York: The Free Press 1974, p. 166.

4. Gale, Richard M. "Human Time" in Richard M. Gale, ed. *The Philosophy of Time.* Garden City, NY: Anchor Books 1967, p. 30.

5. Cottle and Kleinberg. Op. cit., p. 166.

6. Ibid., p. 167.

7. Ibid., p. 168.

8. Priestly, J.B. *Man and Time.* Garden City, NY: Doubleday and Company, 1963, p.160.

9. Ibid., p. 172.

10. Adams, John and Uwe J. Woltemade. "Village Economy in Traditional India: A Simplied Model." *Human Change* 39 (Spring 1970), pp. 49-56.

11. Cottle and Kleinberg. Op. cit., p. 162.

12. Priestly. Op. cit., p. 152.

13. Ibid., p. 158.

14. Ibid.

15. Russell, J. L. "Time in Christian Thought." In J.T. Fraser, ed. *The Voice of Time.* London: Allen Lane, The Penguin Press, 1968, p. 66.

16. Pattaro, Germano. "The Christian Conception of Time." In *Cultures and Time.* Paris: The UNESCO Press, 1976, p. 187.

17. Russell. Op. cit., p. 66.

18. MacGregor, Geddes. *The Hemlock and the Cross: Humanism, Socrates, and Christ.* Philadelphia: J.B. Lippincott, 1963, p. 91.

19. Lodge, George C. *The New American Ideology.* New York: Alfred A. Knopf, 1976, p. 58.

20. Priestly. Op. cit., p. 166.

21. Ibid., p. 64.

22. De Greene, Kenyon B. *Sociotechnical Systems: Factors in Analysis, Design, and Management.* Englewood Cliffs, NJ: Prentice Hall, 1973, p. 207.

23. Max-Neef, Manfred A. *From the Outside Locking in: Experiences in "Barefoot Economics."* Uppsala: Dag Hammarskjold Foundation, 1982, p. 150.

24. Ibid., p. 140.

25. Ibid., p. 150.

26. Priestly. Op. cit., p. 69.

27. Georgescu-Roegen, Nicholas. *Analytical Economics: Issues and Problems.* Cambridge: Harvard University Press, 1966, p. 69.

28. Einstein, Albert and Leopold Infeld. *The Evolution of Physics.* Forge Village, MA: Simon & Schuster, 1938, p. 6.

29. Ornstein, Robert E. *The Psychology of Consciousness.* New York: Harcourt Brace Javanovich, 1977, p. 105.

30. Toffler, Alvin. *Future Shock.* New York: Random House, 1970.

31. Priestly. Op. cit., p. 67.

32. Royce, Josiah. "The Temporal and the Eternal: The Development of Its Philosophical Meaning." In Charles M. Sherover, ed. *The Human Experience of Time: The Development of Its Philosophical Meaning.* New York: New York University Press, 1904 [1975], p. 401.

33. James, William. "The Perception of Time." In Charles M. Sherover, ed. *The Human Experience of Time: The Development of Its Philosophical Meaning.* New York: New York University Press, 1890 [1975], p. 382.

34. Hall, Edward T. *The Silent Language.* Garden City, NY: Doubleday, 1890 [1959], p. 382.

35. Priestly. op. cit., p. 67.

36. Floyd, Keith. "Of Time and the Mind." *Fields* 10 (Winter 1973-74), p. 50.

37. Ibid.

38. Chapin, Jr., F. Stuart. *Human Activity Patterns in the City: Things People Do in Time and Space.* London: John Wiley & Sons, 1974, p. 9.

39. Ibid., p. 10.

40. Ibid.

41. Hayden, F. Gregory with Larry D. Swanson. "Planning Through the Socialization of Property Rights: The Community Reinvestment Act." *Journal of Economic Issues* 14 (June 1980), p. 354.

42. De Greene. Op. cit., p. 61.

43. Ibid.

44. "Real time" is something of a misnomer; a term more descriptive of the concept is "system time."

45. Nelson, Edward A. "A Working Definition of Real-Time Control." P-3089. Santa Monica: The Rand Corporation, 1965, p. 18.

46. Sackman, Harold. *Computers, System Science and Evolving Society: The Challenge of Man-Machine Digital Systems.* New York: John Wiley & Sons, 1967, pp. 223-235.

47. _____. "Futurists on the Future." *Los Angeles Business & Economics* 6 (Summer 1981), p. 23.

48. _____. *Computers, System Science and Evolving Society.* Op. cit., p. 42.

49. Szali, Sandor. "Time and Environment: The Human Use of Time." *The New Hungarian Quarterly* 19 (Summer 1978), pp. 138-139.

50. Max-Neef. Op. cit., p. 139.

51. Szali. Op. cit., p. 134.

52. Ibid.

53. Gurevich, A. J. "Time as a Problem of Cultural History." *Cultures and Time.* Paris: UNESCO, 1976, p. 242.

54. Szali. Op. cit., p. 135.

55. Ayres, Clarence E. *The Theory of Economic Progress.* New York: Schacken Books, 1944 [1962], p. 144.

56. Szali. Op. cit., p. 133.

57. Sackman. *Computers, System Science and Evolving Society.* Op. cit., p. 250.

58. Ibid.

59. Blaunt, J.M. "Space and Process." *Professional Geography* 13 (1961), pp. 1-7, quoted in Edward L. Ullman. "Space and/or Time Opportunities for Substitution and Prediction." *Transactions Institute of British Geographers.* New Series 4 (1979), p. 126.

60. Ullman. Op. cit., p. 127.

61. Parkes, Don and W.D. Wallis. "Graph Theory and the Study of Activity Structure." In Tommy Carlestein, Don Harkes, and Nigel Thrift, eds. *Human Activity and Time Geography.* New York: John Wiley & Sons, 1980, p. 78.

62. Ibid., p. 81.

63. Ibid., p. 77.

64. Ibid., p. 76.

65. Denbigh, K.G. *An Inventive Universe.* London: Hutchinson & Co. Publishers Ltd., 1975, p. 27.

66. Sipfle, David A. "On the Intelligibility of the Epochal Theory of Time." *The Monist* 53 (September 1969), p. 511.

67. Bergson, Henri. "Time as Lived Duration." In Charles M. Sherover, ed. *The Human Experience of Time: The Development of Its Philosophic Meaning.* New York: New York University Press, 1975, p. 226.

68. Parsons, Stephen D. "Time, Expectations, and Subjectivism: Prolegomena to a Dynamic Economics." *Cambridge Journal of Economics* 15 (December 1991), p. 414.

69. Ibid., p. 415.

70. Ibid., p. 419.

71. Whitehead, Alfred. "Two Kinds of Time Relatedness." In Charles M. Sherover, ed. *The Human Experience of Time: The Development of Its Philosophic Meaning.* New York: New York University Press, 1920 [1975], p. 332.

72. Hall, Edward T. *The Dance of Life: The Other Dimension of Time.* Garden City, New York: Anchor Press, 1983, pp. 48-49.

73. _____ . *The Silent Language.* Garden City, New York: Doubleday, 1959, p. 28.

74. Lane, Paul M. and Carol J. Kaufman. "The Standardization of Time." In Robert L. King, ed. *Marketing: Positioning for the 1990s* Proceedings of the 1989 Southern Marketing Association, 1989, p. 4.

75. Juster, F. Thomas and Frank P. Stafford. "The Allocation of Time: Empirical Findings, Behavioral Models, and Problems of Measurement." *The Journal of Economic Literature* 29 (June 1991).

76. Hall. Op. cit. 1983, pp. 53-58.

77. Juster and Stafford. Op. cit., p. 515.

78. Ibid.

79. Ibid.

80. Lane and Kaufman. Op. cit., p. 5.

81. Dewey, John. "Time and Individuality." In Charles M. Serover, ed. *The Human Experience of Time: The Development of Its Philosophic Meaning.* New York: New York University Press, 1940 [1975], p. 423.

82. Jaques, Elliott. *The Form of Time.* London: Heineman Educational Books, 1982, p. 51.

83. Bausor, Randall. "Toward a Historically Dynamic Economics: Examples and Illustrations." *Journal of Post Keynesian Economics* 6 (Spring 1984), p. 360.

84. Ibid.

85. Ibid.

86. Ibid., p. 362.

87. Ibid., p. 371.

88. Norton, Bryan G. "Context and Hierarchy in Aldo Leopold's Theory of Environmental Management." *Ecological Economics* 2 (1990), p. 119.

89. Mitch, William J. and James G. Gosselink. *Wetlands.* New York: Van Nostrand Reinhold Company, Inc., 1986, p. 158.

90. O'Neill, R.V., D.L. DeAngelis, J.B. Warde, and T.F.H. Allen. "*A Hierarchical Concept of Ecosystems.* Princeton, NJ: Princeton University Press, 1986, p. 23.

91. Dewey, Op. cit., p. 423.

92. Whitehead. Op. cit., p. 341.

Chapter 9: Evaluation for Sufficiency: Combining the Social Fabric Matrix and Instrumentalism

1. Dewey, John. *Experience and Nature*. New York: Dover Publications, Inc., 1958, p. 123.

2. Ibid.

3. Norton, Bryan. *Toward Unity Among Environmentalists*. New York: Oxford University Press, 1991, p. 6.

4. Ibid.

5. Daly, Herman E. and John B. Cobb, Jr. *For the Common Good: Redirecting the Economy toward Community, the Environment, and a Sustainable Future*. Boston: Beacon Press, 1989, p. 41.

6. Ostrom, Elinor. *Crafting Institutions for Self-Governing Irrigation Systems*. San Francisco: C.S. Press, 1992.

7. The source of Figure 9-1 is: Kadlec, Robert H. and David E. Hammer. "Modeling Nutrient Behavior in Wetlands." *Ecological Modeling* 40 (March 1988), p. 40.

8. Mitch, William J. and James G. Gosselink. *Wetlands*. New York: Van Nostrand Reinhold Company Inc., 1986, p. 151.

9. Levins, Richard. Evolution in Communities Near Equilbrium." *In Ecology and Evolution of Communities*. Cambridge, MA: Harvard University Press, 1975, pp. 48-49.

10. Ibid., p. 49.

11. Ibid., p. 47.

Chapter 10: The Social Fabric Matrix in a Metapolicymaking Context

1. Dewey, John. *The Public and Its Problems*, 2nd ed. Chicago: The Swallow Press, 1954, p. 135.

2. Lodge, George C. *The New American Ideology*. New York: Alfred A. Knopf, 1980, p. 319.

3. Gellen, Martin. "Institutionalist Economics and the Intellectual Origins of National Planning in the United States." *Journal of Planning, Education and Research* 4 (December 1984), p. 78.

4. Bush, Paul D. "The Theory of Institutional Change." *Journal of Economic Issues* 21 (September 1987).

5. Polanyi, Karl. "The Economy as Instituted Process." In Polanyi, Karl et al. eds. *Trade and Markets in Early Empire*. Glencoe: The Free Press, 1957, p. 249.

6. Dewey. Op. cit., p. 108.

7. Dunford, Richard. "The Suppression of Technology as a Strategy for Controlling Resource Dependence." *Administrative Science Quarterly* 32 (December 1987), pp. 512-525; and Wassily Leontief. "The Choice of Technology." *Scientific American* 252 (June 1985), pp. 37-45.

8. Bollier, David and Joan Claybrook. *"Freedom from Harm: The Civilizing Influence of Health and Environmental Regulation.* Washington, DC: Public Citizen and Democracy Project, 1986, p. 203.

9. Dewey, John. *Reconstruction in Philosophy.* Boston: Beacon Press, 1948, p. xxx.

10. Levin, Samuel M. "John Dewey's Evaluation of Technology." *American Journal of Economics and Sociology* 15 (January 1954), p. 134.

11. Ibid.

12. Tool, Marc R. *The Discretionary Economy: A Normative Theory of Political Economy.* Santa Monica, CA: Goodyear Publishing Co., 1979, p. 296.

13. Quoted in: Lower, Milton D. "The Concept of Technology Within the Institutional Perspective." *Journal of Economic Issues* 21 (September 1987), p. 1161.

14. Ostrander, Gilman M. "Thorstein Veblen." In Gilman M. Ostrander, ed. *Ideas of the Progressive Era.* Clio, MI: Marston Press, 1971, p. 23.

15. Leemans, Arne. "Information as a Factor of Power and Influence." *Knowledge: Creation, Diffusion, Utilization* 7 (September 1986), p. 45.

16. Ibid.

17. Melody, William. "Information: An Emerging Dimension of Institutional Analysis." *Journal of Economic Issues* 21 (September 1987), p. 1337.

18. Blumer, Martin. *The Uses of Social Research: Social Investigations in Public Policy-Making.* London: George Allen & Urwin, 1982, p. 35.

20. Leemans. Op. cit., p. 54.

21. Dewey, John. "The Supremacy of Method." In Milton R. Kinvitz and Gail Kennedy, eds. *The American Progmatists.* New York: Meridian Books, 1960, p. 190.

21. Ratner, Joseph. "Introduction to John Dewey's Philosophy." In Rollo Handy and L.C. Harwood, eds. *Useful Procedures of Inquiry.* Great Barrington, MA: Behavioral Research Council, 1973, p. 23.

22. Street, James. "The Institutionalist Theory of Economic Development." *Journal of Economic Issues* 21 (December 1987), p. 1870.

23. James, William. *Pragmatism: A New Name for Some Old Ways of Thinking.* New York: Longmans, Green and Co., 1907, p. 50.

24. Petr, Jerry. "Fundamentals of an Institutional Perspective on Economic Policy." *Journal of Economic Issues* 18 (March 1984), p. 1.

25. Lodge, George C. *The New American Ideology.* New York: Alfred A. Knopf, 1974, p. 39.

26. Ibid., p. 297.

27. Ibid., p. 95.

28. Ibid., p. 20.

29. Coates, Joseph F. "The Role of Formal Models in Technology Assessment." *Technological Forecasting and Social Change* 9 (1976), p. 145.

30. Ramstad, Yngve. "A Pragmatist's Quest for Holistic Knowledge: The Scientific Methodology of John R. Commons." *Journal of Economic Issues* 20 (December 1986), p. 1097.

31. Lodge. Op. cit., p. 329.

32. Commons, John R. *Institutional Economics: Its Place in Political Economy.* Madison, WI: University of Wisconsin Press, 1935, p. 752.

33. Ibid., p. 749.

34. Quote by Pat Choate in: Rothenberg, Randall. "The Idea Merchant: Pat Choate Sells Economic Policies." *The New York Times Magazine* (May 3, 1987), p. 44.

35. Firth, Raymond. "*Symbols: Public and Private.* Ithica: Cornell University Press, 1973, p. 20.

36. Elder, Charles D. and Roger W. Cobb. *The Political Use of Symbols.* New York: Longman, 1983, p. 21.

37. Firth. Op. cit., p. 427.

38. Elder and Cobb. Op. cit., p. 30.

39. Ibid.

40. Ibid., p. 110.

41. Commons. Op. cit., p. 748.

42. Gellen. Op. cit., p. 78.

INDEX

alternatives
  ecological sets, 191
  evaluate with the SFM, 9, 75, 102-106
  valuation, 64-65
Anderson, Charles, 40
Argyal, Andras, 51
Aristotle, 188
Arnold, Thurman, 109
attitudes
  defined, 81-85
  paradigmatic component, 6
  responses, 82
  signs, 81
  social fabric matrix, 8, 81-85
  system component, 18, 55, 75
  tastes, 83
Ayres, Clarence, 30-32

Bausor, Randall, 177
beliefs (see social beliefs)
Bellah, Robert N., 5
Bergson, Henri-Louis, 158, 172
Berle, Adolf, 210
Bertalanffy, Ludwig von, 59
Bollier, David, 209
Boolean matrix and digraph, 90-94
  graphical clocks, 171-182
Buchanan, James, 214
Bush, Paul Dale, 206

Carter, President Jimmy, 44
Chapin, F. Stuart, 161
Claybrook, Joan, 207
Coase theorem, 15
Cobb-Douglas production function, 19, 84
Cobb, Jr., John, 5, 188, 223
Cobb, Roger, 203
common denominator, 15, 71
Commons, John R., 145, 221
complexity, 1, 6, 54, 59
  democracy, 44
  institutions, 189
  nonequilibrium, 54-55, 195-196
  seek complexity, 27-28, 73

social fabric matrix, 73
consequences, 4, 5, 21, 29, 42
  context, 4
  criteria, 33-43, 187-188
  decreasing options, 188
  democracy, 29-30
  instrumentalism, 29-30, 188
  social fabric matrix, 118, 217-218
cost-benefit, 6, 207
  corporate prices, 15-16
corporations, 15-16, 208
Crafting Institutions, 46, 189
criteria, 4
  context, 33-43, 48, 187-188
  contextual shift, 41-42
  contradiction, 43-44
  cultural values, 76
  democracy, 41-42, 44
  digraph indices, 192-194
  ecological, 5
  normative, 5-6, 8, 35-37
  pluralism, 38-39
  policy evaluation, 37, 194
  primary, 17
  process, 42-43
  scientific, 48
  secondary, 17
  social, 5, 8-9
  technological, 5
cultural values
  criteria, 35-40
  defined, 75-79
  deliveries, 83
  distinction from beliefs, 77-78
  paradigmatic component, 6
  social fabric matrix, 75-79
  system component, 18, 55, 75
  values-beliefs connection, 77, 80
cynicism, 2

Daily Federal Funds Market, 123, 126-132
Daly, Herman, 5, 188
De Greene, Kenyon B., 5, 52
democracy, 41-44
  instrumental approach, 45-46

mandates, 45-47
Denbigh, K.G., 172
Dewey, John, 5, 7, 17
  consequences, 188
  democracy as negative feedback, 58
  instrumentalist ideas, 3
  measurement, 17, 62-63
  scientific inquiry, 47
  technology, 33, 208
  timeliness, 164, 182
Dictionary of Behavioral Science, 81
Discretionary Economy, 5
Dror, Yehezkel, 16

ecological systems
  biodiversity, 11, 103-104
  defined, 84-85
  embedded in sociotechnical, 189
  evaluation, 103-106
  extinction, 192-194
  harmonization in nature, 195
  institutions, 84-85, 189-192
  irreversibility, 193
  living systems, 53, 195-196
  natural goods production, 53
  natural resources, 53
  normative criteria, 19
  options limited, 188-194
  paradigmatic component, 6
  radioactive waste disposal, 109-125
  sink function, 53-54
  social fabric matrix, 8, 84-85, 187-197
  sufficiency, 10-11, 187-197, 192-194
  sustainability, 10-11, 187-188
  system components, 18, 55, 75
  system disruption, 192-194
  technology, 189-190
efficiency
  defined, 20, 64, 195
  judged by consequences, 5
  social criteria, 80
  values and beliefs, 80
Einstein, Albert
  common frame, 1, 74
  intuition, 158
Elder, Charles, 223
Ellul, Jacques, 31-32
Elsner, Wolfram, 97
Encyclopedic Dictionary of Psychology,
    81
environmental policy

organic view, 3
  precautionary, 187
  social fabric matrix, 103-106, 187-197
  sufficiency, 187-197,
  sustainability, 187-188
Environmental Protection: Law and
    Policy, 2
evolution
  clocks and temporal concepts, 165-171
  species, 195-196
  technological, 30-33
  time, 165, 180-181

Federal Deposit Insurance Corporation, 27
Federal Reserve System, 27, 132
Firth, Raymond, 223
Folklore of Capitalism, 109
For the Common Good, 5
Friedman, Milton, 214
Future Shock, 159

Geertz, Clifford, 73
general systems analysis (see systems)
George, Kathryn, 49
Georgescu-Roegen, Nicholas, 158'
Good Society, 5
good society, 1, 5, 14-15
Gould, Stephen Jay, 47
Great Transformation, 5, 50
great transformation, 3-4
growth maximization, 194
GSA, general systems analysis (see
    systems)
Gustrom, Gjessling, 47

Hall, Edward T., 160, 174
Handbook of Social Psychology, 82
Handy, Rollo, 7
holistic, 4, 7, 13, 54, 220, 156
Holmes, Oliver Wendell, 3
Huntington, Samuel P., 76

ideology, 3
  impediments, 195
  whole system, 4
illusion of transparency, 14
Infel, Leopold, 158
institution(s), 2

culture, 77-78
  defined, 79-81
  ecological paradigms, 189-192
  paradigmatic component, 6
  policy, 189-192
  social fabric matrix, 8, 77-78
  system components, 18, 55, 75
  technology, 32
instrumental philosophy
  consequences, 4, 5, 21, 29, 42
  context 187-188
  criteria, 34-43, 187-188
  defined, 3-4, 6-7, 21-30, 187
  democratic policymaking, 7
  development of ideas, 3-4, 7, 21-33
    sufficiency, 187-197
  end-in-view, 4
  experience, 42-43
  ideological impediments, 195-196
  judging by consequences, 7
  problem orientation, 4, 7, 28-29
  social fabric matrix, 187-197
  systems principles, 7
  technology, 30-33
  time analysis, 182-185
  transactional approach to, 7, 33-34

James, William, 3
Johnson, President Lyndon, 221
justice, 2, 14-15

Kahn, Robert L., 51
Katz, Daniel, 51
Kaufman, Carol J., 175
Kerry, Senator John, 34
KISS principle, 27-28
KICK principle, 28

Lane, Paul M., 175
Laszlo, Ervin, 56
Lazwell, Harold, 217
Levins, Richard, 195-196
Lippmann, Walter, 5
Lodge, George, 3-5, 195
Lower, Milton, 209

mandates, 45-47

Mattessich, Richard, 7, 102-103
Max-Neef, Manfred, 157-158, 165
McGuire, William J., 82
McKean, Roland, 64
McMullin, Ernan, 47
measurement
  Dewey on social measurement, 17
  fact/value, 63
  networking and lobbying, 227-229
  primary criteria, 17, 64
  qualification to quantification, 17, 63
  secondary criteria, 17, 65
  social 17
  social indicators, 62-66, 223
Melody, William, 211
Menand, Louis, 3
metapolicymaking, 203-231
  advocacy, 225-229
  diagrammatic overview of, 204-205
  ideology, 218-220
  instrumental philosophy, 216-218
  levels and phases, 203-213
  participatory democracy, 217-218
  problem definitiion
  problems and consequences,
    216-217
    problem definition, 221-222
    program budgeting, 229-230
    select program, 224-225
  social fabric matrix context, 11, 222
  sociotechnical change, 230-231
  transactional processes, 7
  theory
    creation of integrated systems,
      213-214
    diffusion of knowledge, 214
    linkage to theory and operations, 215,
      230-231
    utilization of knowledge, 214-215
methodology, 51
  science is a policy area, 47-49
models
  common, 1
  complex, 73
  deterministic, 3, 59
  integrated process, 5, 18, 38
  interactional, 23-27
  logical positivism, 47
  mechanistic, 54
  one-dimensional systems, 52
  rates and levels, 163
  reductionist, 54

self-actional, 23-27
sustainable, 191-192
transactional, 22-28
Myrdal, Gunnar, 209

National Environmental Policy Act of
  1969, 3
Neale, Walter, 79
neoclassical economics
  common denominator, 15
  corporate prices, 15-16
  category mistake, 16
  hedonism, 15
  illusion of transparency, 14
  objectivity, 43-44
  time analysis, 155-158, 175-176
  utility, 13-15
normative criteria, 19, 80, 102-103,
  192-194
networks
  social fabric matrix digraphs, 91, 93,
    96-97, 114, 129-132, 136, 138,
    140-141, 173, 179, 181
  social fabric matrix *ithink* digraphs,
    121, 124-125, 130-132
  social fabric matrix, 97
  timeliness, 183
  defined, 18, 97
New American Ideology, 5
noncommon denominator, 86-88
non-equilibrium, 2, 52, 86-87

objectivity, 42-43
one best way
  not viable, 43-45
Open Market Desk, 131
Ornstein, Robert E., 158
Ostrom, Eleanor, 46, 189

paradigm
  general, 6
  paradigmatic shift, 2-3
Pareto optimality, 15
Patee, Howard, 56
Peirce, Charles, 3, 7, 33, 40
pessimism, 2
Petr, Jerry, 214
Plato, 44
Polanyi, Karl, 3, 5, 49, 162, 214

policy, 199-205, 209-226
  analysis, 4, 195-196
  decreasing options, 188
  evaluation, 10-11, 102-106, 187-197
  paradigm, 6
  SFM, 11, 75, 102-106
  science, 47-49
  sufficiency, 187-197
  systems, 187
policymaking
  advocacy, 220-224
  delivery criteria, 194
  inappropriate paradigms
    bureaucratic, 206-207
    pseudo-strategic, 206-208
    scholarly kings, 208-209
  strategy, 199-205, 209-226
  tactics, 199-205, 209-226
  timeliness, 162-165
problem
  context, 4
  orientation, 4, 28-29
  social construction, 28-29
Public and Its Problems, 5
Quest for Certainty, 47

Ramstad, Yngve, 217
Rosen, Robert, 54
Royce, Josiah, 159
rules, 20, 22, 25, 45-46
  belief criteria in contracts, 111-117

Sackman, Harold, 164
Schumacher, E.F., 13, 31-32, 156
Schumpeter, Joseph, 31
  delivery criteria, 194
Semiotics
  attitudes, 81
  corporate prices, 15
  real-world experience, 40
  signs, 34, 81
Seven Cultures of Capitalism, 146
SFM (see social fabric matrix)
Shackle, G.L.S., 172
Siffre, Michel, 159
Smith, Adam, 195
social beliefs
  contract terms, 109, 114-117
  defined, 79-81
  deliveries, 83

normative criteria, 19, 35-40
paradigmatic component, 6, 17
primary system controls, 56
social fabric matrix, 79-81
system component, 18, 55, 75
systems, 3-4
social fabric matrix (SFM)
applied cases (see social fabric matrix
   case studies)
attitudes, 81-83
Boolean matrix/digraph, 90-94
cellular information, 89-90
complexity, 73, 187
component integration, 85-86
cultural values, 79
defined, 5, 8-9, 73-75, 85-88, 106-108,
   118, 203
delivery concept, 85, 192-194, 196-197
ecological systems, 84-85
evaluation, 102-106
flow concept, 85-86
illustrations of, 9
indicators, 192-194
institutions, 79-81
instrumentalism, 187-197
ithink, 9, 119, 121-125, 128-132
metapolicymaking, 11, 187-197, 222
noncommon denominator, 86-88
post-policy, 9
pre-policy, 9
principle of association, 73
principle of model exchange, 73
process matrix, 86-87
social beliefs, 79-81
spreadsheets, 93-94
sufficiency 187-197
system analysis, 94-106, 187
system components, 18, 55
technology, 84
time analysis, 184-185
timeliness, 145, 184
social fabric matrix case studies, 110-143
Central Interstate Compact, 109-125
   Bechtel, 110-118
   beliefs as contract terms, 111-114
   cell deliveries, 114-117
   context, 118
   contract analysis, 114-119
   cost, 113-119
   institutions, 110-120
   conclusions, 142-143
   Daily Federal Funds Market, 123-132

   cell deliveries, 126-128
   Federal Reserve System digraph, 132
   findings, 130-131
   institutions, 128-130
   Open Market Desk System digraph,
   131
   ithink networks, 109, 121, 124-126,
   130-132
livelihood strategies in India, 133-134
   cell deliveries, 135
   conclusions, 133-134
   livelihood strategies digraph, 136
Platte River surface water assessment,
   134-142
   cell deliveries, 138-139
   regulatory institutions and water
     users digraph, 141
   regulatory institutions digraph, 140
   water management digraph, 138
social indicators, 7-8, 16-17, 223
abstract measures, 8
common denominator, 71-72, 102
consequences, 65
creation, 16-18
criteria indices, 192-194
design, 62-66
instrumentalism, 62
multiple criteria 61, 102-103
monitoring, 65
normative interest, 62
primary criteria, 17, 64-70
problem orientation, 61-62
qualification to quantification, 17, 63
relationship, 65
requirement, 65
role of criteria, 8
secondary criteria, 17, 64
Sociotechnical Systems, 5
Summers, Lawrence, 39
Swaney, James A., 54
systems
alternative paths, 55, 98-99
complexity, 51, 54
constituents and components, 52, 55,
   100
control, 55-56, 99
defined, 51, 54-55, 94-95, 98-99
deliveries, 57, 100
differentiation and elaboration, 59, 101
equifinality, 55, 101-102
evaluation, 60, 102-106, 162-165,
   182-185

external description, 52-53, 58
feedback
    negative, 52, 57, 100
        defined, 58
    positive, 52, 57, 100
        defined, 58
    flows, 57, 100
general systems analysis (GSA)
    defined, 7, 51-60
hierarchy, 56-57, 100
internal description, 52-53
non-equilibrium, 2, 52
nonisomorphic, 54, 97-98
norms, 102-103
openness, 52-54
    defined, 52-53, 95-97
    inputs from environment, 52, 96-97
rates and levels, 163
real time, 59-60, 101-102
regulation, 55-56, 99
relationships, 51, 55
requirements, 55
sequences, 57, 100
social fabric matrix, 94-108
state, 52
subsystems, 95-97
technology, 165-171
timeliness, 162-165
Szalai, Sandor, 165

technological society, 1, 2, 30-32
technology
    change without progress, 210-211
    combination, 30, 84
    defined, 84
    institutional guidance, 32
    instrumentalism, 30-32, 37
    normative criteria, 19
    paradigmatic component, 6
    policymaking
        corporate control, 206-208, 211-212
        impediments in use of, 210-211
        limiting options, 188
    positive feedback, 58
    social fabric matrix, 8, 84
    system components, 18, 55, 75
    system control, 56
    time and clocks, 165-171, 184
Theory of the Leisure Class, 61
time
    analysis, 156-161

baseball, 156-157
Christian, 149-151
clock, 10, 156-158,167-170
cultural, 147-148
cyclical, 148-149
defined, 10, 145-149, 155-156
evolutionary, 180-181
graphical clocks, 171-182
instrumental analysis, 182-185
Kantian system, 59
neoclassical discounting, 151-154
Newtonian, 59
passing, 146, 153
processes, 165, 171-172, 174-181,
    178-181
psychological, 158, 160
real time, 59-60, 101-102, 163-165,
    168, 172-173, 178
social time, 101, 147, 171-172, 182
societal construct, 60
subjective, 158-160
technological evolution, 165-169, 184
temporal conditions, 10
traditional, 10
Western, 149-156
timeliness
    appropriate concept, 10, 145
    defined, 10, 60, 145, 161
    planning decisions, 161
    social fabric matrix, 145
    system planning, 161-162, 164
transactional
    approach to science, 21-28
    complexity, 26
    context, 34
    defined 25-26
    instrumentalism, 22-28
Toffler, Alvin, 159
Tool, Marc, 5, 42
Tugwell, Rexford, 210

U.S. General Accounting Office, 70-71
U.S. Supreme Court, 27, 179
utilitarian theories, 14-15
utility, 6, 13-15, 158

values (see cultural values)
Veblen, Thorstein, 33, 61

Warfield, John, 73
Weisbrod, Burton, 39-40
Westbrook, Robert, 4
Western values, 77
wetlands, 19, 84-85, 189-192
Whitehead, Alfred, 84, 183, 188

Young, Iris, 14

Printed in the United States
133585LV00003B/175/A

9 780387 293691